Let Light Into Your Heart
with Colour and Sound

Companion volume to her best-selling book

The Healing Tones of Crystal Bowls

by Renee Brodie

Published by
Aroma Art Ltd.
Vancouver, Canada

Editor: Neall Calvert
Book design and typography: Fiona Raven
Watercolours: Renee Brodie © 2001
Drawings: Paul Brodie © 2001
Photographs: Jack Brodie and Crystal Tones © 2001

Published by Aroma Art Ltd.
548 - 48th Street, Delta, BC V4M 2N3 CANADA
Telephone: (604) 943-6689 Fax: (604) 943-6660
Web Site: http://www.angelfire.com/ca/crystalbowl
Email: renee-brodie@telus.net

Printed in Canada by Hemlock Printers Ltd., Burnaby, B.C.

Canadian Cataloguing in Publication Data
Brodie, Renee, 1920–
Let light into your heart with colour and sound
A companion to The healing tones of crystal bowls.
Includes bibliographical references.
ISBN 0-9680790-1-6
1. Crystal bowls—Therapeutic use. 2. Color—Therapeutic use. 3. Sound—Therapeutic use. 4. Vibration—Therapeutic use. I. Title. II. Title: The healing tones of crystal bowls.
RZ414.6.B76 2001 615.8'3 C2001-900913-5

NOTE: I am a teacher and researcher, and report findings that I feel may be beneficial to others. I have no connection with the medical profession and therefore do not advise anyone to cease any medical treatments, nor to stop seeing their doctor. Any information given in this book is not intended to be taken as a replacement for medical advice. Any person with a condition requiring medical attention should consult a qualified practitioner or therapist.

Also by the author:
Life Energies, published 1984
The Healing Tones of Crystal Bowls, ISBN 0-9680790-0-8, published 1996, third printing 2000

Acknowledgements

So many, many thanks to those kind ones who have each made such a wonderful contribution and given great encouragement towards getting this book "on the road."

The list includes my loved husband, JACK, on hand to help in many ways with great humour; our son, PAUL, for his beautiful drawings, and our daughter, FIONA, for not only her wonderful design expertise, but also her colouring arrangements, her musical help, exciting ideas with graphics and putting it all together with skill, understanding and insight; and NEALL CALVERT for his very precise editing.

Thanks also to Merry Bakker, Uta Buckingham, Debbie Cottle, Diane Fay, Alice Friend, Helene Harris, Ph.D, Janet Hobbs, Barbara Huff, William Jones, Jan Linch, Marie-Louise Lacy, Betina Lindsey, Julia Miller, Petra Robinson, Norman Shealy, MD, Patricia Tarling, Paul Utz, and once again to JONATHAN GOLDMAN for his informative and much appreciated Foreword.

I give deep thanks to all of you, and God Bless.

Renee Brodie

Dedicated . . .

. . . with deep love to Those Shining Ones who so inspired me with Their writings, Their patience and Their love;

—to my dear husband JACK for his loving support, encouragement, editing, wonderful humour and above all, his strength to me;

—to our children, PAUL and FIONA, who are my dearest teachers;

—to BONNIE and LAMBERT WARE for their kind and wonderful, spontaneous generosity;

—and to the grateful memory of STUART PAWSON, who put me "on track" printing my first book and was so kind and helpful.

"[The] amount of dense food we need for nourishment is really very small, but the amount of food and nourishment which we absorb through breathing, sound and colour is remarkably high. . . . [A]s we see this in relation to the body, we must also see it as a nutritional factor to the Soul, for the Soul's food is also colour, sound, form and scent. It has been said that the Cosmic essence flows from the heart to the glands. This is quite true. Colour therapy is a small indication that this is possible, and though our range of colours on the third dimension is still crude we use colour therapy as reliable means of re-energizing depleted fields. Each of us has an entirely different key of vibrations . . . So our colour fields, our colour zones, energies and fragrance are like our fingerprints—identical to ourselves. But there are certain Universal colour fields or nourishments to which we can apply the science of occult thinking, and this the old Alchemists used to distil, from the flowers the essence of health, and from nature the aroma of healing."

Ronald Beesley
The Creative Ethers

FOREWORD

"And the Lord said: Let there be Light!"

—*Genesis, The Old Testament*

From ancient times, knowledge of the use of sound and light to heal and transform has found its way to us through many of the sacred and mystical texts. Today, through continued honouring of the ancient sources as well as exploration of the new sacred sciences, we are seeing a tremendous emergence of awareness of the power of sound and light as tools for transformation.

The importance of light, sound and other vibrational modalities cannot be understated. We are living in extraordinary times, with increasing shifts of frequency—changes in vibration that can assist us on both a personal and a planetary level. These frequency shifts are quite powerful, and as such, can also create imbalances. It is my belief that through understanding and using these vibrational modalities, we can help create adjustments to ourselves in order to resonate in peace and harmony.

From my perspective, we are unique vibratory beings—each one of us part of the Divine One and yet each slightly different. I teach that what resonates, or vibrates harmoniously, with one person may not necessarily resonate with another. Even in traditional medicine this is true. Penicillin is an antibiotic that will assist a large percentage of the population. But a small percentage of people who take the drug will have an allergic reaction. Not everything is good for everyone. This holds true for the vibrational healing modalities such as sound and light. Thus, it is impossible to state accurately that this frequency will have this specific effect for everyone. We can speculate on a generic response, but not suppose the effect will be universal. We need to remember our own vibrational uniqueness.

It is my experience that we all have the ability to shift our frequencies and change our vibrational rate. Frequency shifting allows us to experience fluidity and not become locked into static waveforms. This ability to shift frequencies seems to have been embodied by the great spiritual masters who have walked this planet, and is something to which we can all aspire. As we continue in our evolutionary path towards higher consciousness, it is most important that we open ourselves to the potentials inherent in this concept of frequency shifting. This concept may be a key to our conscious acceleration on this planet. Thus the idea that one's root chakra, for example, resonates to the color red and the keynote of "C" may become a self-limiting system after enough experience working with the chakras. Nevertheless, initial knowledge of this information is quite necessary, for it gives us a common basis from which to ultimately emerge into the realm of frequency shifting!

An understanding of systems of light and sound is most important—particularly for those who may be new to the vibrational medicine playing field. Before one can lose the training wheels on a bicycle, it's necessary to learn how to ride. The same holds true for using light, sound and other vibrational modalities. Before one can truly learn to shift frequencies, it's important to have knowledge of what others perceive of as the effects of light and sound. It gives us a wonderful basis from which to explore this field. Much information about these subjects can be found in Renee Brodie's book.

Let Light covers an abundance of information on many different subjects related to the area of sound and light; Renee encapsulates the wisdom of numerous teachers. I applaud her for gathering this material into such a coherent package. You'll learn about the chakras, the aura, the effects of colour and sound, the Indigo Children, Crystal Bowls and other information that will assist with creating a more balanced and harmonious life. If you have not resonated with these concepts before, you'll find this book most useful. If your awareness of the subjects covered within this book is deep and vast, you will still find useful gems of information.

There is much beauty expressed in these pages, coming from a soul whose heartfelt journey into Light and Sound is both interesting and enjoyable to read. I trust you will resonate with Renee Brodie and, through this book, bring more Light into your lives.

Harmonically Yours,
Jonathan Goldman

Table of Contents

Foreword iv
Introduction x

1 Light 11–17

The Great Unrest Within 11
The Giving of Oneself 12
Meditation: The Inner Sanctuary 13
Light is a Living Spirit 14
Light Treatment in 1901 15
On Becoming the Light 15
Clearing the Channels 16
Meditation: Becoming the Light 17

2 The Wheel of Life 18–20

The Journey Begins with One Step 19
Meditation: Dawn of a New Day 20

3 Birth 21–33

Conscious Conception 21
Cultivating a Beautiful Atmosphere for the Incoming Life 23
We Choose Our Birth 24
The Incoming Soul 25
Abortion 27
Birth 29
Spring and Birth of the New 29
Let the Joybells Ring Out 30
Case History with Crystal Bowls 30
"Tucking In" the Newborn's Auric Field 31
A Mother's Love 31
The Nursery 32
Negative Patterns 32
Meditation: Springtime Cottage 33

4 Childhood 34–40

Making Childhood Meaningful 34
The Indigo Children 35
Help your Children through Colour 38
The Zen of Parenting 39
Meditation: The Simple Daisy 40

5 Teenage 41–47

Within the Light 41
The Teenager 42
Black 43
White and Black 45
White 45
Waves of Light 46
Meditation: Into the Light 47

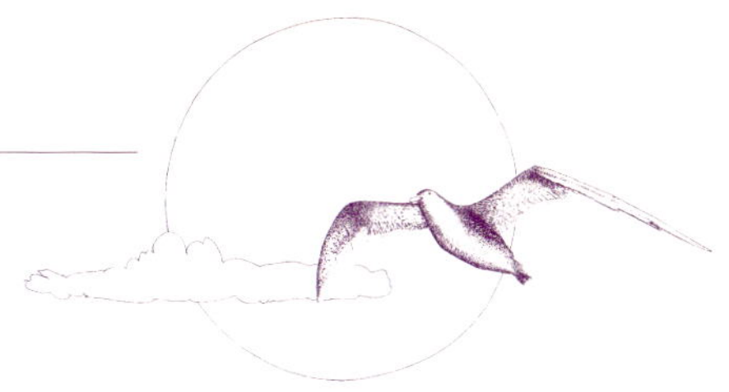

6 *Your Beautiful Auric Field* 48–52

Your Beautiful Aura 48
The Auric Field 49
The Simplicity of Life 51
Meditation: The Calm Sea 52

7 *The Philosophy of the Chakras* 53–59

The Chakras 53
Newly Awakened Chakras 55
The Truth about Chakras 56
Never Underestimate the Healing Power of Sound 58
The Gossamer-like Patterns of Life 58
Meditation: Journey to a Rainbow 59

8 *Sound* 60–76

Music Hath Charms 60
Sound 61
Keynote—Own Note—Soul Note 63
Your Own (Personal) Note 65
What then is Sound Healing? 66
Vowel Sounds 66
What is Toning? 67
The Tuning Fork 68
Searching for your Soul Song? 69
Sounding your very Own Tone 71
What is Your Number and Colour? 72
Sharry Edwards—Sound Healing Pioneer 74
Music Played Publicly Everywhere 75
Rock and Pop Music 75
Meditation: Armchair Healing 76

9 *The Pure Tones of Crystal Bowls* 77–81

The Practitioner Crystal Bowl 77
New Developments in Crystal Bowls 79
Meditation: On the Wings of the Wind 81

10 *Colours of Life* 82–89

So What is Colour? 82
All Life is Colour 84
Colour Therapy is Soul Therapy 85
Soul Colours—Another View 85
Working with Complementary Sounds and Colours 86
History of Spectro-Chrome 87
Treatment by Light 88
A Clear Blue Sky 88
Meditation: The Blue Heavens 89

11 Red—The Blazer of the Trail 90–96

On Purifying the Bodies 91
Chakra #1—Root 92
Grounding and Cleansing 93
Case Histories with Crystal Bowls 93
Pink 94
Music for the Root Chakra 94
Whither Thou Goest—There Go I Too! 95
Suggested Music for Anger 95
Meditation: Journey in Colour 96

12 Orange—The Spirit of Health 97–102

As We Think, So We Are 97
The Orange Ray 99
Chakra #2—Sacral 100
Music for the Sacral Chakra 100
Case Histories with Crystal Bowls 101
Meditation: A New Day 102

13 Yellow—The Spirit of Wisdom 103–111

On Walking in the Path of the Sun 103
The Yellow Ray of Wisdom 104
Chakra #3—Solar Plexus 105
Yellow, Red, Blue or Green Filters for Dyslexia 107
The Panic Button and Fear 108
Decrystallizing Fear 109
Tinted Lenses Create Problem 109
The Clear Yellow Beam 110
Music for the Solar Plexus Chakra 110
Case Histories with Crystal Bowls 110
Meditation: Golden Energy 111

14 Green—Ray of Balance & Harmony 112–121

Your Inner Healer 112
The Green Ray 113
Chakra #4—Heart 114
Love and Colour 116
Music for the Heart Chakra 116
A Strong Heart Centre 117
Case Histories with Crystal Bowls 117
Friendship 118
Companionship 118
Holy Matrimony 119
Down Memory Lane 120
Meditation: The Rose of Your Heart 121

15 Turquoise—Ray of Tranquillity 122–127

On Being Gentle and Kind in Whatever You Do 122
The Turquoise Ray 123
Chakra #5—Thymus—The Higher Heart 124
Case Histories with Crystal Bowls 124

Affirmations for the Thymus Chakra 125
Music for the Thymus Chakra 125
The Blooming of the Ethereal Rose 126
Meditation: The Bridge 127

16 *Blue—Ray of Peace & Serenity* 128–135

The Light We Carry Within 128
Blue—The Greatest Healing Ray 129
Chakra #6—Throat 130
The Throat Centre—Gateway to Consciousness 130
Case History with Syntonics 132
Music for the Throat Chakra 133
Case Histories with Crystal Bowls 133
Speak Your Truth, Speak Your Peace 134
Meditation: The Bluebird of Happiness 135

17 *Indigo—Ray of the Aquarian Age* 136–142

The Magic of the Eyes 136
Indigo—The Soul Colour 137
Chakra #7—Third Eye 137
Music for the Third Eye Chakra 137
Gravity/Magnetic Balancing 140
Case Histories with Crystal Bowls 141
Meditation: Still, Still Waters 142

18 *Violet—Ray of Spirit* 143–151

"To Thine Own Self Be True" 143
Chakra #8—Crown 144
A Life for A Life 144
Balancing the Astral/Emotional Body with Crystal Bowls 145
The Process of Death (or Rebirth) 146
Music for Rebirth 148
Resting Assured that Life Goes On 149
Case Histories with Crystal Bowls 149
Time Marches On 150
Music for the Crown Chakra 150
Meditation: The Mists of Time 151

19 *Magenta—A Master Ray* 152–157

Connecting Daily with our Higher Self 152
The Magenta Ray 153
Case Histories with Crystal Bowls 154
Music for the Astral Chakra 154
Our Magnificent Soul Body 155
How Do People Recognize the Soul Part of You? 156
Releasing the Cobwebs 156
Meditation: The Temple of Your Star 157

Notes 158
Bibliography 160

Introduction

This companion book to *The Healing Tones of Crystal Bowls* is the product of much love, Light and travail!

The essence of the book concerns the Light we kindle within our hearts, much needed now in this brand-new century, against a background of Birth, Life and Rebirth, together with more ideas on Colour and Sound.

More and more people are now being drawn to the sounds of Crystal Bowls and the incredible results experienced in balancing the body, mind and spirit.

As before, all inspirational writings are in italics.

This writing is offered to you, dear Reader, with great love, and may the Light enfold and enfill you while you listen to the words deep in your heart.

Renee Brodie
Tsawwassen, B.C.
March 2001

Chapter One

Light

LET YOUR LIGHT SO SHINE that not only will your works be seen but the Path you have blazed in your incarnating experiences will live on as a great glow, as a centre of heat and warmth, affection, Light, love and giving . . . that the world will be so much richer that you have loved and lived and served and fought in its midst—helped to produce the heaven on earth, because you have in some small way contributed towards its future creation.

—Anonymous

The Great Unrest Within

The soul reflects the Light within outwards, that is to say, whatever Light, Love and Peace there is within each soul is reflected through the eyes and also through the feelings deep within the heart. We cannot feel without showing that feeling through the eyes, and also the skin, for in Joy and Love the skin glows, not just in the face, but all over the body and we emit this beautiful radiance all around us.

Think of a most beautiful, lovely tree. How do you feel when you gaze upon it, how do you feel when you stand beneath its outstretched arms? Very uplifted? Often filled with a mysterious wonder and a joyous feeling inside? Remember, the great energies of the Spirit are everywhere, using every means possible to touch everyone.

So the very next time you go out into Nature, the very next time you look into someone's eyes (the mirror of their soul), see what you see, feel in your heart—are they emitting Joy, Love? Is the radiance of Nature in all its glory touching you? Are you so moved that you can hardly express yourself, your delight, your Joy, except perhaps exclaim, or want to dance—or jump for Joy—because you are so touched?

This very feeling then is the one we endeavour, through you, to evoke within each and every soul, that they can be impelled a little further on their Journey through life, that we can touch the very core of their Being, their soul quality, and spark their life for them like an encouragement—almost like a perfume—that their unfoldment can be more in tune with their soul qualities, and therefore reach out to others, so uplifting them too.

This is soul quality—it touches many hearts, not just one; it flows onwards and each is blessed in their very own way to carry forward the golden Christ Light magically on their special Journey through life.

The Giving of Oneself

There is so much thought around the idea of "giving"—what is it that one can give? How much does one give of self? Does it have to be all give? Can one keep part for oneself? Does it have to be recognized, appreciated? Does one talk about it?

Down through the ages there have always been those whose dedication to service for others was remarkable, with no thought for self, simply being there for others' needs. But do we know in everything there must be a true balance? We must take time for ourselves first, to nourish, to care for our wellbeing and to be well, otherwise how can we give of our best? How can we hope to be clearer channels? For are we not cleansing continually?

The whole idea now must be one of balance, with the knowledge that to give is to receive. In the act of giving we are actually receiving too, particularly in the healing process as we channel the beautiful energies, like an empty vessel; first the energies travel through us and we can partake of them. It is so true that the more we give, the more energies are available through us—it is a Divine Law. In our daily lives then, we can be more aware of this wonderful order, firstly to offer ourselves as a channel, allow whatever is needed to happen—the important things will unfold first and all else shall fall into place.

Monitor yourself—stand back as it were—and see yourself as your day unfolds. Watch for the opportunities, be open to being available where necessary. But this also means being aware firstly of one's own needs, for sick people can be very demanding and there are fewer excellent channels for healing. Say "NO" if you need to.

In other ways, it is the quality and tone of your voice which heals, and often just one word will suffice—it will be greatly remembered and make all the difference. Best of all is just Being—it is the radiance poured through you touching another without words or even recognition. This is so healing for many. Let us then, each of us, as channels, walk through life, aware and awake, observing, caring, giving our love, giving the God-given energies to heal, to balance, to bring Joy and Peace in our own special corner of this majestic Planet Earth, and in the giving be so grateful.

We have not yet assimilated this century's most far-reaching discovery: namely, that energy, or Light, is the principle underlying all manifestation. In other words, we have not yet grasped that we, our bodies and our minds, are Light.

—Georg Feuerstein
from the Foreword to
Drawing the Light from Within: Keys to Awaken Your Creative Power
by Judith Cornell

Meditation: The Inner Sanctuary

See yourself walking up a grassy slope. . . . It's a beautiful day,
and whichever way you look there is a beautiful, breathtaking view. . . .
As you climb the hill easily, you see ahead a small white building. . . .
Its door is opened invitingly, so enter. . . .

Find yourself in a Sanctuary . . . very peaceful and quiet . . .
simply furnished, as YOU like it . . . perhaps a thick carpet or rug on the floor
(in the colour you like) . . . a picture you love on a wall . . . some lovely flowers
on a small table. . . . One whole wall is window . . . there's a garden outside,
and beyond . . . a magnificent view.

Sit on your kind of chair, or cushion. . . . Enjoy the peace and quiet
and stillness in this little room . . . for it is your very own Inner Sanctuary.
Here you can come any time you need to . . . for quiet, renewal, contemplation . . .
and even to commune with a Shining One, who will come and help you
with your questions if you call. . . .

Stay awhile in the Silence and the Peace. . . .

LIGHT IS A LIVING SPIRIT

Solar power . . . is not merely an energy on the physical plane. Light is a living spirit which comes from the sun and which has a direct contact with our spirit. . . . To the degree that light is able to come into the human soul, it is reflected as intelligence, light, beauty, nobility and strength.

. . . If only you had an idea of the Higher Beings living in the sun who send the rays to us! They look down on us, smiling at times, yet we are unaware of their presence. I realize that it is difficult for people to recognize that light is more than a physical vibration, that it is in fact a living spirit; but if they are closed to this idea, they will not be able to receive all the sun's blessings.

Sometimes our Light goes out but is blown again into flame by an encounter with another human being. Each of us owes deepest thanks to those who have rekindled this inner Light.
—Albert Schweitzer

You must arrange your life so that light will play an increasingly large part in it. If you are to benefit from it, you must become conscious of its presence in everything. . . . Breathe with the knowledge that by breathing you are able to draw light into you and you prepare the best conditions for receiving heavenly light, the Spirit of God. . . .

In the morning when you look at the sun, think that these rays which are coming to you are living beings who can help you resolve all the problems of the day ahead. . . . Instead of merely looking at it, drink and eat the sun, visualizing this light spreading throughout all the cells of your body, purifying, strengthening and giving them life. . . . Put everything else aside, just study the rays of the sun. . . . If only you had an idea what power, richness, clarity, purity and intelligence each ray of sun contains!

. . . Spiritual Beings live in each ray, each manifesting differently according to their colours of red, blue, green, yellow, etc., and for this reason it is very important for the disciple to learn how to work with the light.

. . . If you succeed in concentrating on light, feeling it as an ocean which vibrates, throbs and quivers, where all is peace, happiness and joy, you will also begin to feel that this light is perfume, and music, that cosmic music of the spheres, the song of all that exists in the universe.

Nothing is more noble, glorious and powerful than working with the light.[1]

—Omraam Mikhaël Aïvanhov

Light Treatment in 1901

Long before there were medical lasers, another kind of light was used to heal the sick: sunlight. In the nineteenth century a Danish professor named Niels Finsen discovered that eruptive skin infections could be treated by dosing patients with steady levels of red light. Soon smallpox sufferers were being taken to treatment rooms where midday sun streamed through red windows and red drapes. Later Finsen found that parasitic and microbial skin conditions could benefit from the violet waves lying at the other end of the light spectrum. Finsen would project white light through hollow lenses filled with blue water and shine the beam on diseased skin patches. At one treatment center in Copenhagen, four patients could be treated at once.

—*Discover Magazine*, November 1988

On Becoming the Light

From the very Heavens, if we are aware, there comes an enormous emanation of spiritual energies, all-enfolding, encompassing, uplifting, touching each and every soul on this beautiful Planet. And if we are aware, however we feel or see or sense, then can we use this wonderfully uplifting energy for good—to embrace it, enfill ourselves in every way, and as there is such an abundance of it, the more we use it, the more we are given. We can use it to help others in a healing way, and because it is channeled THROUGH us, then as we use it we are ourselves healed, and more comes to use on others.

This all-embracing beautiful Light energy is there for us to use. It is so packed with energy, we cannot only energize ourselves but know that we have become like a Light, soft at first. Then as we grow spiritually, this Light becomes stronger and stronger, lighting our heart centre, filling it to overflowing until It reaches the centres above and below the heart, ignites them as it were, and in turn, the ones above and below are fuelled and lit, a marvelous Christmas-tree–like effect—effervescing, bubbling, dancing energies touching others who come within the orbit of this special Light.

There are many special Beings whose Light so shines, and when we meet them we know them. Even if we cannot put words to them, we feel uplifted, joyous, good, filled with encouragement, Love and good feelings to help us on our daily pathway.

You are one of the Light Brethren. Nurture this Light, this heavenly feeling, and let your Light shine very brightly each day. Nourish it, polish it, take care of it, let it flow through you in an unbroken stream. Be that Light for others and fulfill your mission on Mother Earth with Joy and understanding!

Clearing the Channels

Those of you who are drawn to these words are all channels for the Light, for the teachings and the healings, the energy pulsating through you, and in this work we must be totally aware of the implications of its specialness, especially at this crucial time when the beautiful Mother Earth is experiencing a rebirth in multifaceted ways, as we too are so doing.

In order to be a clear channel for the Light, we must first understand that is exactly why we chose to incarnate at this time, to be here, to help in every way possible, often very unobtrusively, but where it's necessary—unsung, quietly, above all with Love and knowledge.

In order to prepare for this we have learned in many lives how to respond to cries for help. We have learned how to properly care for our bodies, in all aspects—physically, mentally, emotionally and spiritually—and we have also learned how to just BE, that we are where we are for a reason, to accept that, to carry on our lives normally, without show, but paying attention to detail. That means health, caring, emotions, Joy and wellbeing for those around.

Then, having created a beautiful home and atmosphere of wonderful peace within, we are there for all those who are guided to us by the Shining Ones. It may be for simply one word; it may be for counselling, for healing or for the comfort of being in your home for awhile to experience the peace. But always there is the help at hand given THROUGH you by the Shining Ones, who look for the Light shining within your hearts, and therefore the aura—then can they direct people to you.

In all this, then, we see clearly that we must daily be on the alert, firstly for ourselves, to comfort our bodies, to exercise, to feed nutritionally, to rest enough, and to MAKE time for quietness each day. These are priorities. Only then can we be a fit vehicle to channel the Light. The Shining Ones always watch over us. They help by "nudging" us with intuition. They encompass us with their radiance, filling us to overflowing, and when we give of that "effervescence" it is always replaced twofold—let us remember that!

Henceforth then, each day let us be SO thankful, so open to the Loved Ones, that we can with gratitude fulfill our mission, however lowly, for all adds up to the whole—and go about our lives rejoicing, filled with this Holy Light which spills over to all who touch us, and in so doing they too can shine, feel appreciated and encouraged to emulate our example. For we need many more clear channels to do the work of moving into the New Era of Love and Light for everyone. Then shall Mother Earth rejoice, and be healed herself.

Meditation: Becoming the Light

When you enter the stillness and feel the quiet enfold you . . .
feel the harmony, warmth and Love fill your very Being. . . .
See before you a great expanse of land,
a vast stretch of waving cornfields in the Sun. . . .

And towards the horizon, as you watch, there appears like a column,
stretching from the yellow corn into the Heavens, a beautiful band—
vertical—of rainbow colours . . . soft pastel shades from palest lilac to pink
and blue, with all the golden tones between. . . .
As you gaze upon its tonal beauty, it begins to expand to each side,
getting wider and wider, and the colours paler and paler. . . .

Then in the centre you become aware of a beautiful, glorious figure of Golden Light—
the Light is so bright you can hardly see the figure, but you sense it is there. . . .
You feel its emanations towards you—of warmth and Love,
deep compassion . . . a sense of great strength and understanding. . . .

The Light grows even brighter . . . stronger . . . and shines over you—
like outstretched arms—until it completely engulfs you . . .
filling every cell in your Being with a great Peace and Love and Healing. . . .

YOU ARE THE LIGHT. . . .

Chapter Two

The Wheel of Life

Whatever you can do or dream you can, begin it.
Boldness has genius, magic and power in it!

—Goethe

In our own way each day we perform a miniature Wheel of Life. We do certain things, we turn the Wheel ourselves by our thoughts and deeds, like a performance, a ritual each day almost the same. When we consider the Wheel of Life, we realize how absolutely magnificent it is—birth, life and rebirth. For some, the three most important events in human life would be Birth, Marriage and Death; we like to arrange them as Birth, Life and Rebirth.

We elect to come to Earth to gain more experience in certain ways. We choose our vehicles (parents) for what they can offer us in the way of experience. We live our lives, some more exuberantly than others, hopefully learning at each step of the way, and finally, when our chosen time comes (however subconsciously!), we are reborn to our prebirth home—to relax, review, and maybe later, start again to learn yet more about ourselves.

If we can see life in this way, life becomes more interesting, perhaps more accepting, understanding. We can also see others in a better Light, struggling towards the Light, get to know ourselves more intimately, and therefore make the Pathway so much easier, more beautiful, more peaceful in every way.

Let us then, together, walk through these pages with more understanding for ourselves and so for others, with Light in our hearts.

The Journey Begins with One Step

In the beginning we always hope to set aside earthly worries and woes, to see a far broader horizon, to see where the trail leads in our life, to see into the distance where a great Light beckons us to we know not where.

However, one step at a time. Take the first step, then you are encouraged all the way, without any hesitation. Not always can we see further ahead, but always a sure step in the right direction if we listen—then is our next step assured. So we encourage you to step out with confidence, leave behind that which you already know and be very open to new learning.

Always there will come one onto your Pathway. It will be guidance—in some form or another: an encouraging word, a smile, a hug even—helping you to love yourself, see your own inner beauty, to appreciate yourself, and to go forward with Joy in your heart, for is not the world so beautiful?

Mother Earth gives to us, all the time, of her energy, her perfume, all her delights for the eye—the colours, the warmth of the Sun, the blue of the sky, the gentle touch of the breeze, so many things for which we can be so thankful. Let us fill ourselves with fresh air, and march onward with lightness in our step to try once more on our Pathway, which is so Lighted if we will take the time to look and listen, and in the doing feel so very blessed. . . .

And as we travel our Pathway, we can stand back as it were and watch the unfolding of a wonderful pattern, teaching above all, but also loving each step along the way. We can know we are watched over, that if we need help, we have only to ask. Above all, it is "Thy Will be done, not mine, dear God"—our prayer to start each new day.

> *If the mind and body are to be well, you must begin by curing the soul.*
>
> *—Plato*

Meditation: Dawn of a New Day

In the stillness of the misty morning air before dawn, find yourself sitting
on a log on the beach. The world is so still . . . hardly a sound
but for the gentle lapping of the water. . . . You sit there, still . . .
engulfed in a beautiful warmth of your thoughts of wonder
and love for all Nature . . . the sheer beauty of it . . .
of the anticipated dawn of a new day. . . . So many lovely things can happen . . .
so much can eradicate past pains by letting go . . .
seeing them fade into the distance of the hills when the mist clears . . .
which it begins to do as you gaze outwardly.

Live and let live. . . . You are well, and grateful to be so. . . .
There is much to do with your life, but a step at a time is enough. . . .
Here in this waiting space is all perspective . . . is a release of guilt and fears . . .
is love for yourself, a beautiful Being . . . is knowing that the Radiance
being poured into you is not only healing but can be passed on to others . . .
just by Being . . . as simply as sitting right here. . . .

SUDDENLY, awareness comes. Mists begin to clear. The rich red of the dawn
begins to rise from behind the tips of the mountains across the sea. . . .
The richness fills the whole space . . . splashes into the water,
reflects towards you. You watch—fascinated—
until the colour gradually brightens to a warm orange . . . a shining gold . . . the SUN!
It rises slowly as you watch in amazement at its simplicity . . . its splendour . . .
filling the whole world with warmth, hope and encouragement
for the brand-new day—that we too, can fill it
with our radiance, hopes and Light of Peace. . . .

Chapter Three

Birth

Conscious Conception

Let us now take an in-depth look at Birth, at the many facets of one of the most important events in life, beginning with a view of Conscious Conception.

Alice Friend is a member of the White Eagle Lodge, which has a worldwide spiritual "family." She writes about conscious conception in a wonderfully heart-warming way in their bimonthly publication, *Stella Polaris*:

> As the Age of Aquarius is influencing us more and more, people are becoming more conscious, more aware of the spiritual influences in their lives. These can be anything from being aware of God's beauty in washing the dishes, mowing the lawn, scrubbing the floor, to the more attractive and easily distinguishable miracles of life such as being peaceful in a special garden, or watching magically the birth and growth of a flower or human being one loves.
>
> The birth process is such a miraculous and sensitive experience. Maybe we should say "re-birth" process, because that is more to the point. It starts with the seed, and when the seed splits and comes out of itself, a Being is born and starts to grow. So the planting of the seed is as important as the nurturing of it.
>
> Take the seed of a flower. If you love to garden, you are careful where and how you plant the seed. You make sure all the conditions are favourable like the soil, the time of year, the moisture, etc. It should be the same with the conception (or the planting of the seed) of a human being. Many women find themselves pregnant without being aware of the magical moment of conception. Is not then conscious conception as important as conscious birth? Or in other words, being truly awake to the very secret of life itself?
>
> The souls of human beings never die, they live on, with God (or in the Light) forever. Only the outer garment, or the body, dies and when it does, it dissolves into the earth, setting free the pure Light form of the soul, the true essence of our God-selves. In this pure state, whether in a body or not, we love, we feel, we think, and we are aware of life with or without the physical density surrounding it. Coming from the world of Spirit into the physical is not an easy journey for the soul, and often, coming into the density of the physical body, we lose certain sensitive qualities. They seem absorbed in matter and forgotten. The spirit takes second place to the material.
>
> Our bodies are given to us by the God-essence as a Temple for the Spirit and we must learn to use them as such. We must

take care of our Temples then and learn how to use them as God intended, for we are given a body to teach us lessons not possible to learn in pure spirit form. Souls we have loved never leave us. Often we meet more whom we are to love and from whom to learn. Therefore, within a family, we would rarely find those who have not been together in past lives. What a lovely gift this is; knowing it consciously, and living it, makes life much richer and fuller. So, when conceiving a child, the two partners are, in essence, vehicles of God for His Plan to take form. What a special opportunity to be given! The act of making love itself is a magical gift, a special gift two human beings can share and enjoy in the bliss of their love. So when a couple conceives a child through that act, our physical bodies merge with spirit in the presence of the divine love, so we are one with God, the Universal Father Mother. Being fully aware and in love with life (Life being God presence) we are awakened in spirit and mind to the secrets of the Universe!

Conscious conception to me means being fully awake to being used by God for a vehicle of His love. This is so for both the man and the woman. The soul of the child has long before been drawn to the parents according to the law of Karma.

The child-spirit is within the aura of the parents long before the conception actually takes place. A sensitive woman can recognize this and often "hear" within herself the soul communications with her. The magical moments of meditation can often bring this through very clearly.

So the child is conceived through an act of divine Presence and love. Then the nine months of growth and slowly coming into incarnation in the physical. Suppose we were conceived and then born without the nine months' gestation time—our nervous systems would short-circuit! We need time to adjust from being pure Light in spirit form, to being surrounded in the denser, physical body. Then the actual birth itself, sometimes a painful process for both mother and child—the newborn child of the earth takes time to adjust to its new body and world. The parent must understand this and surround the child in love, forgetting selfish desires. In doing this, the sensitive parent is rewarded on a deep level, one hundred-fold. What a miracle life really is, what a joy to be truly awake within this miracle and know we have been given such a special opportunity to live out the will of God in all we do.[2]

Recommended Book
Life Before Life, by Helen Wambach, Ph.D. (New York: Bantam Books 1981).

Cultivating a Beautiful Atmosphere for the Incoming Life

A thought then about colour. It is extremely important for the expectant mother to wear lighter colours—to allow the Light through to the growing foetus. Remember that dress materials act as filters so that the body receives colour, and if we are wearing the wrong colours it affects us and we feel uncomfortable. Pregnant women, it is said—and seen—radiate Pink in their auric field!

The soul entity hovers around the Mother, and finally takes possession of the growing foetus usually at about four-and-a-half months.

> In humans the ear is the first sensory organ to develop. It is fully functional four-and-a-half months before we are born. In the womb the sounds we hear—our mother's breathing, her heartbeat and especially her voice—stimulate our brain and fire electrical charges into our cortex. It is the nourishment provided by these electrical charges that enhances our mental function and spurs the proper development of the brain and the central nervous system. Because the ear is an active channel to the brain and nervous system, it acts as a battery, constantly charging and stimulating them both. This stimulus is critical not only to our body's ability to grow properly and develop muscle tone and co-ordination, but to our ability to both hear and listen, and ultimately learn and understand.[3]
>
> —Ron Minson, M.D.

> As a mother conceives and begins to build a foetus for the soul attracted into incarnation through her, it is very important to cultivate a beautiful atmosphere for the incoming life. The home environment that is clean, lovely, with bright colors and paintings will allow Light to permeate the household, through the windows and through the consciousness of the householders.
>
> Again, lovely music can help to prepare the way for the incoming child, and it heightens the atmosphere in the home.[4]
>
> —Hal Lingerman

Earl Mindell's Recipe for Morning Sickness:
To alleviate nausea and discomfort caused by hormonal changes, drink a cup of ginger root tea or take a ginger root tablet or capsule first thing in the morning. Throughout the day, sip ginger ale or peppermint tea.

Welcoming Music
Here are some calming, welcoming pieces that will help the infant to enter Earth's density in joy and warmth:

- *Brahms—Lullaby*
- *Debussy—Clair de Lune*
- *J. S. Bach—Air on a G String*
- *Mozart—Piano Concerto No. 21 (2nd movement)*
- *Wagner—Evening Star (from Tannhäuser)*
- *Lullaby from the Womb—Dr. Hajime Murooka (compilation CD—12 tracks).*[4]

—list by Hal Lingerman

We Choose Our Birth

Jan Linch, in her book *Starchild*, says:

> Before you are born, you see your life review in front of you like a video. It all seems so easy when you are up there vibrating at a much higher frequency, but when you fragment [come to Earth] things are not quite so clear.
>
> You choose your life.
> You choose your parents.
> You choose your name.
> You choose your mission.
> You choose your difficulties.
>
> All that you choose is carefully mapped out whilst you are still within your Soul Group Energy. This is your free choice before you are born but the choice of *how* you decide to deal with each situation your life comes up against and what you can learn from each one is *yours*. You choose your parents, often to repeat their patterns, or perhaps to learn from them, maybe even how not to be. In choosing your birth you also choose the experiences that go with that birthing, right from conception. You may choose a water birth or a natural birth. You may choose Caesarian or to be induced. You may choose a traumatic birth linked to your fears or labels.
>
> Your date of birth is carefully chosen, for part of who you are is within your birth chart, not just your star sign. The sun indicates the date of birth in your chart. However, the moon and other planets in your chart also have very strong influences. . . .
>
> You map out your life with such precision that throughout your entire life, you actually attract everything in the course of it that you have chosen to deal with or work through. You will find that, throughout your life, you will keep repeating patterns until you deal with the issues behind them. Many of the faults and judgments you find in other people are actually a reminder of your own, for the faults you see in others are often, in fact, a mirror image of your own.
>
> You chose your parents and family, not always for the first time either. Everything you choose in life is a learning process whereby you can grow to achieve that ultimate balance.[5]

Delivery by Caesarean Section

The rate of Caesarean delivery in the U.S. now approaches 25 percent of all births. Caesareans are more expensive, and in many cases more dangerous, than vaginal delivery. Most C-sections are done as "defensive medicine" to avoid perceived threats of malpractice actions.[6]

—Dr. Andrew Weil's *Medical Practices to Avoid #11*

Becoming a New Zealand Midwife

On December 24, 2000, CBC Vancouver's early morning show was interviewing Canadians working abroad. Among them was a young lady who, being unable to become a midwife in British Columbia, went to New Zealand and qualified there after attending university. She now takes complete charge of the patient, diagnosing and prescribing drugs as required, and having hospital privileges when necessary.

Surely our nurses and midwives should be trained to be less reliant on doctors, and be capable of operating small clinics in isolated areas where doctors are not available? Given the opportunity, it is certain the midwives will measure up.

The Incoming Soul

Barbara Ann Brennan, in her book *Hands of Light*, states:

> . . . [T]he incoming soul often enters and leaves the body through the crown chakra as it begins working on opening the root chakra to grow roots into the physical plane. The root chakra looks like a very narrow funnel, and the crown chakra looks like a very wide funnel at this stage. The other chakras look like small shallow Chinese tea cups with a narrow line of energy leading back into the body to the spine. (See Figure 1).
>
> As a baby fixes his attention on an object in the physical plane, the aura tenses and brightens, especially around the head. Then, as his attention fades, the aura fades in colour; however, it retains some of the experience in the form of colour in the aura. Each experience adds a little colour to the aura and enhances its individuality. Thus, the work of aura building is also going on and continues in this way throughout life so that all of one's life experiences can be found there.
>
> After birth there remains a strong energy connection between the mother and child. This connection is sometimes referred to as the germ plasma. It is strongest between mother and child at birth and will remain there throughout

The general field of a baby is amorphous, formless and has a bluish or grayish colour.

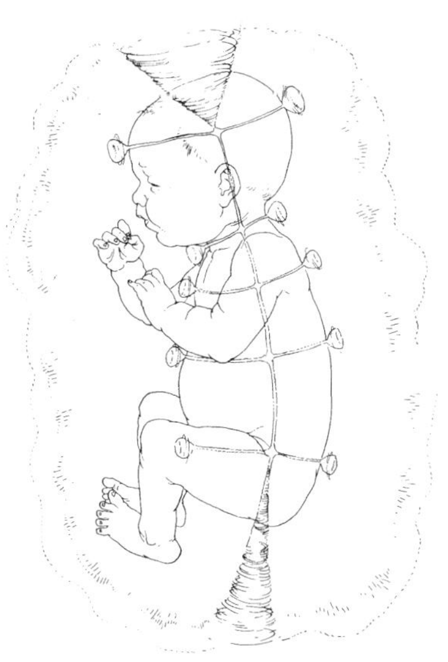

Figure 1

Massage Delivers Babies

Deep-tissue Massage therapy is showing promise as a treatment for some types of infertility. The technique was discovered accidentally by Larry and Belinda Wurn, clinical co-directors of Clear Passage Therapies in Florida. Several of their female patients, previously diagnosed as infertile, became pregnant after two to three months of weekly massage treatments. Statistics show that this effect is more than a coincidence. Of eight women treated in a preliminary study, 50 percent have given birth to healthy, full-term babies. Conventional fertility treatments like in vitro fertilization have a combined success rate of only 19.6 percent.

"We were initially surprised by our apparent success rate," Larry Wurn said, "then we became fascinated." The Wurns, with a team of researchers, are now conducting a controlled clinical study of the treatment's effectiveness on women with blocked fallopian tubes. If you have been diagnosed with two blocked tubes and want to participate in this study, call Clear Passage Therapies at (904) 332-8000 for more information.

life, although it will not be so pronounced as the child grows. This psychic umbilicus is the connection through which children remain in contact with their parents over the years. Many times one is aware of traumatic experiences the other is having, although there may be great distance between them on the physical level.

The field of the child is entirely open and vulnerable to the atmosphere in which he lives. Whether things are "in the open" or not, the child senses what is going on between his parents. The child constantly reacts to his energetic environment in a manner consistent with his temperament. He may have vague fears, fantasies, tantrums or illness.

The child's chakras are all open in the sense that there is no protective film over them which screens out the incoming psychic influences. This makes the child very vulnerable and impressionable. Thus, even though the chakras are not developed like those of an adult, and the energy that comes into them is experienced in a vague way, it still goes right into the field of the child, and the child must deal with it in some way. (See Figure 2 to compare adult and child's chakra).

At around the age of seven, a protective screen is formed over the chakra openings that filters out a lot of the incoming influences from the universal energy field. Thus, the child is no longer as vulnerable as before. This stage can be seen as a child grows and individuates. It is near the time of the dawning of reason. . . .

The process of slowly awakening to the physical world continues after birth. The baby sleeps frequently during this time; the soul occupies its higher energy bodies. It leaves the physical and etheric bodies disengaged and allows them to be very busy doing the work of body building. . . .

They still have some awareness in the spiritual world, and I have seen them struggling to let go of spiritual playmates and parent figures and to transfer affections to the new parents. The newborns I have observed have very wide open crown chakras (Figure 1). They are struggling to squeeze themselves into the confinements of the tiny body of a baby. . . . [I]n their higher bodies, they appear many times to be spirits of about twelve feet tall. They go through an enormous struggle in opening the lower root chakra and connecting to the earth.[7]

Adult's Chakra

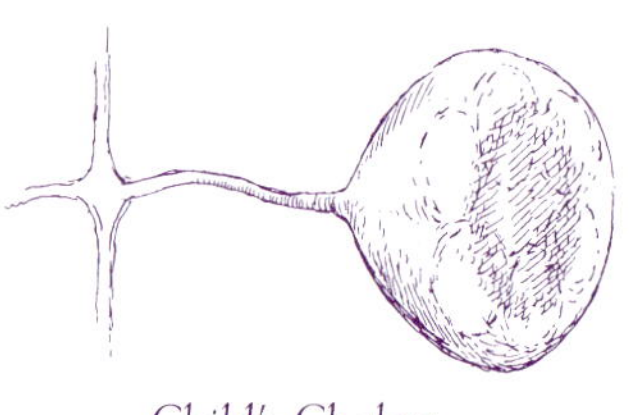

Child's Chakra

Figure 2

Abortion

In great Light we see loved Souls approaching their Earth Vehicle so that they may come to Earth for certain reasons . . . to be here at a certain time of fulfillment, above all to help Mother Earth through the knowledge they can bring with them; also to evolve more themselves.

So the approach is made, a soul can be known to hover around the mother for awhile before conception takes place, and then, when it does, to make a final decision whether to stay or not, and for the mother to decide too.

It is felt that this decision can be made safely within the first few weeks after conception—to be or not to be, to stay or not to stay and also whether to be with this person as a channel for birth. There are many considerations, but above all, it is the LOVE which cements a beginning relationship. This is felt acutely by the incoming soul—a great yearning for love and true understanding. Finally, the great decision is made by both parties; perhaps the mother is simply not ready for this Passage of Light and all it entails.

Sometimes a pact is made on a subconscious level: "Not now, please . . . but do try again later when I am more ready. I send you Home again with great love and ask for your understanding."

After an abortion there is great need for care, for a great cleansing inwardly, to acknowledge that it is fine, that you can offer an opportunity later if necessary when you feel more ready, but that meantime, you are grateful for the experience, you have more knowing, you love your body and will take greater care of it, will acknowledge that you are fit and well, and your body is the Temple of your Spirit, so you are caring for it with love and understanding, without any regrets, guilt—only a greater feeling of Love, and that you do need more rest and Joy, that everything has a time, place and a knowing.

Allow peace to fill your whole Being and know that the body is a great work of art and beauty. Be thankful, for your Angel is there to guard and guide you onwards, enfolding you with great Light, and that Soul can now move on to find a more appropriate vehicle and take the Light within too.

Dr. Gladys McGarey discusses abortion in her book *Born to Live*:

> The spontaneous abortion is no real psychological or ethical tug-of-war, but when it comes to an induced end to a pregnancy, there are literally thousands of well-meaning men and women on both sides of the dividing line. And I was torn between the two warring factions.
>
> As the concept of the continuity of life grew in its reality in my mind, and the pieces started to fall into place—how we have no beginning and no end as spiritual creatures; how we choose our parents; how we may see what lies ahead, no matter what it may be; how experiences are for our growth and understanding; then I began to feel a bit better about abortion when it was really needed. The idea of reincarnation helped.
>
> But I still had a problem with women who wanted an abortion because it was convenient, or because the pregnancy was embarrassing or threatening the

stability of a marriage. These things didn't make sense to me.

. . . [H]owever, a new day dawned in my consciousness. . . . I can see that it is frequently reasonable, understandable and the "right" thing for a woman to do. . . . [It] dawned with a story a patient told me some time ago. . . . This mother had a four-year-old daughter whom she would take out to lunch occasionally. . . .[when talking] the child would shift from one subject to another, when Dorothy suddenly said, "The last time I was a little girl I had a different mommy!" And she started talking in a different language. . . .

The magic moment seemed over, but then Dorothy continued, "But that wasn't the last time. Last time when I was four inches long and in your tummy, Daddy wasn't ready to marry you yet, so I went away. But then, I came back." Her eyes lost that faraway look, and she was chatting again about four-year-old matters.

Mother was silent. No one but her husband, the doctor and she knew this, but she *had* become pregnant about two years before she and her husband were ready to get married. When she was four months pregnant, she decided to have an abortion. . . . When the two of them did get married and were ready to have their first child, the same entity made its appearance. . . . It does throw a new light on abortion, doesn't it? [After reading this passage, a friend told Renee Brodie: "My sister had a similar experience. Her young son asked, 'Why didn't you let me come the first time?'"] Every experience is for the individuals involved, and all abortions do not have this kind of circumstance surrounding them, naturally, but I could see how things were different than I at first thought.

If the entity really understands what is happening, if there is a reality to consciousness in all spheres of life, and if a couple makes the decision to have an abortion in an understanding, thoughtful manner—with the feeling that somehow this is really the right thing to do—then somehow, I think, there is communication with the Divine, and the outcome is bound to be a constructive, learning event. . . .

When one of these situations arises now, I ask the young woman to take time to write down all of her reasons and feelings *for* having an abortion and the reasons and feelings *against* having one. Now in as unemotional a way as possible, make the best decision she can make for all concerned. Then prayerfully go about the business of life, either getting the abortion done or not.

It has surprised me how many of the girls, after going through this process, have decided to keep the pregnancy.

I now always tell the mother to communicate with the entity who is hovering, waiting to be born. "Tell that individual what is happening, and why it is happening. This makes things more in accord with life patterns and needs for all those involved. . . ."

Is abortion, then, something to take lightly?[8]

—Gladys McGarey, M.D.

Birth

We are at the Gates of Heaven, it is so LIGHT—there is Lightness all around, and we are preparing to journey to Earth, to take with us our Light, our colours, and also our experiences enfolded within so as to be able to use them for our Journey on Earth . . . to help others in their understanding.

As we approach the idea of emerging through the darkness of the birth channel into the Light, we ask fervently of the Light Beings assisting that we shall be received with great love and all-encompassing Light, gentleness and compassion and above all JOY. We ask that we shall not forget our Heavenly Home, but to bring with us "samples" of that feeling of belonging, of gentleness and above all, great understanding of who we are, so that our chosen parents will understand, be touched by our thoughts and handle us with a gentleness and even greater forethought because at this time and Age, we come with special intent, to help Mother Earth, to be the gentleness added to firmness of intent, in fact, to be a great Light at this special time.

Let the birth process, then, gently unfold, helped by the midwife and doctor, and the presence of the caring father. Let there be gentle music of our choice; let the sound permeate every cell of each, calling forth the best and touching our heartstrings to evoke great kindliness, Love, Joy and above all understanding, that the whole of the birth room shall be bathed in pink and blue, that the passage to Earth for this particular soul shall be Lighted . . . and they can be welcomed with great Peace and Love for a perfect "landing."

Spring and Birth of the New

Listen to sweet birdsong, alive in the trees, with the sunlight streaming through the leaves, a light spring green—so delicate—and above, the blue, blue sky. Let us walk in Nature, breathe in the scented aroma of new growth, fresh earth lightly touched by rain in the night, the warmth of the sun's rays bringing forth the scent of the Mother Earth's rich loam.

And as we walk we are reminded of newness, new birth, new leaves, new fresh aromas—and of all the beautiful souls returning to Earth in this Heavenly Springtime, coming with renewed vigour and old knowledge to inspire us, to move us onwards in our quest for peace, and to encourage us that all is not lost—these old souls in new young bodies who bring great Love and inspiring endeavours to enrich the Earth and our lives too.

Let us send love to each newborn child of God, that they are greeted with great Love and Joy, enfolded in the Light, that their "landing" is peaceful and easy, that they shall be greatly cared for, wrapped in Love and Light for their journey onwards.

It is Springtime for them too, in their lives—for the moment, so much to see, do and learn, so much need of encouragement. Let us enfold and keep them in the Golden Christ Light so that their learning is joyous, fruitful and filled with DeLight.

Let the Joybells Ring Out

On the clear mountain air on a glorious day there is heard a bell, ringing very clearly across the valley. It is sweet, it pierces the heart, it is full of joyous sound which touches us, saying, "Come and be with me; come and rest your weary Being for a while. Refresh yourself; be my guest; leave all your problems on the shore and climb to the highest heights. Be with me and just melt into the sound from my heart centre; be happy and relaxed and think only of the Sun, the glorious Sun. Let it fill your whole Being with glorious golden Light, and then allow it to overflow through the very pores of your skin to all around, scintillating Light into the mountain air. Here is breathtaking beauty, no cares—all have receded and Joy has taken over. You are whole in every way.

"Today is your Birthday, a day in which to be filled and then reborn, ready and willing for the next year to unfold in all its many glorious facets, and you are SO STRONG, you know without any doubt at all that you CAN cope, you can see beyond the challenges that come; everything is so very clear to you at this moment in time. You feel so great, grateful too, for always there is help, in often the most unexpected ways, and always at the right time for you."

Be happy with these thoughts, then—wrap them around you. Be still in the Light in the mountain air, and in the heart of that very stillness give thanks and love to yourself, you are such a beautiful Being, you are so loved. Wrap that sunlight around you, encompass yourself in the stillness, and all is as preparation to return to your home in the world and resume as before, but with a great difference within, a greater space for Love and kindness, compassion and clear-seeing, so that your gentle essence and clarity of being can be used not only for yourself but for other Beings seeking along the Pathway of Life.

Be Joyous, let your Joybells ring out, piercing the air, allowing clouds to clear and the sunlight to pour through at last!

Case History with Crystal Bowls

I recently worked with a woman who had just delivered her third baby, and because of different reasons the delivery was traumatic for both mother and child.

The child couldn't stop crying, which made the mother feel very guilty and nervous which upset the whole family. I suggested she bring the baby so I could work with them through the auric field. . . . I met a very angry little baby (soul) who had lost her way, didn't want to arrive and be here. So I tuned in for help and put the baby in between the crystal bowls . . . took a very soft striker/wand and sounded the bowls, meanwhile, silently talking to the soul telling her that I understood how difficult it was to enter into such a small body and into restriction, also encouraging her to make the connection so that life on Earth could become a journey in joy and creativity instead of repetition of pain and restriction. The mother, watching this, was still very upset, but saw and experienced how the baby first stopped crying, and after calming down gave us the most wonderful smile!

—Annet Hoeijmans-Boon, colour and sound therapist, Eindhoven, Holland.

"Tucking In" the Newborn's Auric Field

Dr. Gladys McGarey writes:

> One of the stops Bill [husband] and I made on the 1969 World Tour was in England where our group met with Ronald Beesley [founder of the College of Psycho-Therapeutics in Kent, now known as the Centre for New Directions]. He talked with us . . . drawing on the black-board with coloured chalk his concept of the auras he saw when he looked at people. In most of his drawings . . . the colours joined each other above the head in a very symmetrical fashion. But with others, there was sort of an overlapping at the top.
>
> "Why don't the auras meet at the top in some people?" I asked.
>
> "These are the people whose souls were not tucked in properly when they were born," he answered. He expressed in words what I had instinctively known, but had been unable to put into words myself. I was really thrilled.
>
> It means that, when a baby is born, it is our duty in officiating at such occasions to "tuck in" the aura—the way I look at that, it simply means that we have to bless that child real good! We need to handle that new arrival with gentleness and the awareness that this is a special moment in their consciousness—a transition from spirit into the material dimension. We need to be aware enough that we help in drawing the life energies, the forces of life itself, together so that the baby can adapt and function normally in this world of ours.[9]

A Mother's Love

The greatest love that comes to mind at first is that of a mother, so warm, so encompassing, so at-one with the child. It is beautifully nurturing, enfolding, and as that soul grows, with such an understanding it blossoms before the Mother's eyes, grows into something very lovely. She sees not too much wrong, and what a blessing that is. This is an ideal love, as it should—or could—be, such an enfolding nourishment acting as a real base for true growth.

But there is an even more wonderful Love that abounds, that of the Great Spirit—call it the Father of us all. Within the bounds of this great Love there is peace, security, unfoldment, nurturing and it is offered in so many ways, mysterious ways, often by the touch of a stranger's hand, by a look from beautiful eyes filled with love, by a chord in Nature—the call of a bird: the beauty of sound—which touches another chord within us. Even the love of a child has such beauty and can evoke feelings within us to draw out the best.

So wherever we are, daily in our lives, let us be so mindful of where we can touch where it is needed, craved even. Let us be very aware, on tiptoe, to reach for the highest motives, always touched by more loving compassion, and as we can reach out, so can the messenger for us be at hand at the right moment, uplifting us to the Great Spirit within. For there it is, right in the heart, ready to shine brightly and warm—someone sent by Spirit to us.

The Nursery

Often expectant parents design the child's room before birth, and it is a good idea to make sure the colours chosen will give your child a wonderful beginning for life on Earth.

A baby needs to be enfolded with soft, light colours, and if we could see their aura it would have all pastel shades, soft tones of pink and blue, peach, yellow or a creamy colour. These tones are very soothing and comforting, as Rudolf Steiner taught earlier this century—colour and shape can have a very beneficial effect on our physical and mental development, and in his Waldorf Schools there were special colours for each Grade (see Childhood chapter, page 34).

It is suggested to not dress young babies in brighter shades, as their aura cannot take it—it makes them overactive, restless and sometimes aggressive.

Strong Colours

Strong, bright colors have the effect of shocking the baby's inner vibrations, which could make the infant unsettled and restless. Bright, intense colors such as primary red, yellow, and orange can stop a child sleeping well, as well as causing him or her to cry. Strong contrasting colors and bold patterns are also likely to be overstimulating, so for a small infant choose soft tones of yellow or cream, peach or pink, which radiate peace and are emotionally soothing and comforting.[11]

—Suzy Chiazzari

Negative Patterns

> The colors of the child's bedclothes and bedding have a direct effect on his or her physical and psychological well-being because the subtle color vibrations permeate the aura while the child is asleep. Unless the child is well balanced and outgoing, be careful not to have strong, large, or geometric patterns on drapes, blinds, or walls. These shapes can send off jarring vibrations. Distraught parents have often come for advice regarding restless children who suffer from recurring dreams and nightmares. Often the child has a quilt or pillow decorated with black, red, purple, and other bright colors, covered with large cars, trains, or comic figures. Removing these negative vibrational patterns from the room is the best thing to help your child get a good night's sleep.[10]
>
> —Suzy Chiazzari

There is a cycle of changing colors in our aura through the different stages in our lives. These inner colours are reflected in our changing color preferences.[12]

—Suzy Chiazzari

What is the human body but a constellation of the same powers that formed the stars in the sky.

—Paracelsus

Meditation: Springtime Cottage

It is springtime; the air is fresh, warm. There is expectancy in the air. . . .
You feel light-hearted and walk with a buoyancy down a lane.
There are hedges on each side and you hear birdsong—
delightful sounds of Nature. . . .

Soon you come to an opening in the hedge and turn left, down a beautiful pathway
leading to a cottage. . . . This is the garden. . . . On each side of the path
are flowers of every hue: pink . . . orange . . . golden yellow . . .
blue . . . green leaves and stems. What a glorious rainbow—
and the perfume . . . !

As you walk slowly on the grassy path, you notice the small white cottage
has a beautiful doorway. See the colour of the door. . . . It is opened invitingly. . . .
Perhaps you feel like entering, for there is a warm welcome awaiting you.
Love and Peace fill the whole place.
There is Joy and laughter and understanding. . . .
Feel the warmth of the hearth . . .
sense the fragrance of Love and the joys of Spring. . . .

Enjoy your own peaceful, beautiful Sanctuary
and be refreshed for the next step on Life's Journey. . . .

Chapter Four

Childhood

Making Childhood Meaningful

Childhood can be such a joyous, carefree time of growth and becoming, especially if the parents are loving, kind and very aware of all the necessary ingredients for encouragement, care, discipline and nurturing of their child.

This needs a rounded, well-balanced "diet" from both parents, of constant reinforcement of your love, along with wise counselling, of having a feeling of knowing what exactly is needed for the moment, to teach the child to be independent, to be loving and kind to others, to not expect, for expectations are "killing" and may never happen. Above all, to respect elders—if a child has this response, then he/she too will be respected in turn. And to teach that to give is to receive—a very important rule in life, that they be allowed to help in the home and be praised for it.

These are all positive things, with no hints of the negative—express EVERYTHING positively—and where there are siblings, to give to each of them of your time (most important); let them feel wanted (needed is a better word) but not dependent. This makes for self-reliance which is SO important. Above all, love each of your children and tell them so every single day. This is so nurturing (and necessary) to feel wanted and needed, brings a glow to the centre, for children are very sensitive. They "know" inside themselves just how you feel and how honest you are.

Make their lives meaningful and encourage all you can; it will be a brilliant Light within them that is nurtured.

Waldorf Schools paint classrooms to correspond with the "soul mood" of the children at particular stages of development. Colourful surroundings have a remarkable effect not only on challenged children.

Grade 1 - Medium Pink
Grade 2 - Orange Curtains, wood-panelled room
Grade 3 - Yellow
Grade 4 - Green or Light Yellow/Green
Grade 5 - Mint Green
Grade 6 - Pale Blue
Grade 7 - Darker Blue

In the early grades, they are surrounded by warmth to help them grow; the colours become gradually cooler as the intellectual faculties develop. There are strong magnetic colours in the early grades, later green for more study and blue for more judgement.

The Indigo Children

There is a remarkable book called *The Indigo Children,* by Lee Carroll and Jan Tober, that offers encouragement and help to parents of these beautiful children who have come to Earth beginning about the early 1980s to help both Mother Earth and their parents. I want to quote extensively from this book:

> What is an Indigo Child? And why do we call them *Indigo*? . . . [A]n Indigo Child is one who displays a new and unusual set of psychological attributes and shows a pattern of behaviour generally undocumented before. . . . Here are ten of the most common traits of Indigo Children:
>
> 1. They come into the world with a feeling of royalty (and often act like it).
> 2. They have a feeling of "deserving to be here" and are surprised when others don't share that.
> 3. Self-worth is not a big issue. They often tell the parents "who they are."
> 4. They have difficulty with absolute authority (authority without explanation or choice).
> 5. They simply will not do certain things: for example, waiting in line is difficult for them.
> 6. They get frustrated with systems that are ritual-oriented and don't require creative thought.
> 7. They often see better ways of doing things, both at home and in school, which makes them seem like "system busters" (nonconforming to any system).
> 8. They seem anti-social unless they are with their own kind . . . they often turn inward, feeling like no other human understands them. School is often extremely difficult for them socially.
> 9. They will not respond to "guilt" discipline ("Wait till your father gets home and finds out what you did").
> 10. They are not shy in letting you know what they need.

In the same book, Nancy Ann Tappe says:

> My statement is that 90 percent of the children under ten are Indigos. . . . I first started noticing it in 1982. . . . I call them Indigos because that's the color I "see." . . . At about 26, 27 you're going to see a big switch in the Indigo Children . . . their purpose will be here.

All parents, particularly parents of Indigos, could benefit greatly from the following books as well:

Back in Control—How to Get Your Children to Behave, by Gregory Bodenhamer (New York, Fireside, 1988).

The Life You Were Born to Live—A Guide to Finding Your Life Purpose, by Dan Millman (Tiburon, CA, H. J. Kramer, 1993).

Tolbert McCarroll, a modern-day monk who draws from a wealth of life experiences, writes:

> *When you guide a child you help him develop his strength. You teach him not to ask you to solve a problem that he could face himself. The spoiled child does not want there to be any problems. If there is an adult who caters to him the child never develops his strength and goes through life holding tight to his mediocre images of existence.*[13]

There are four different types:

1. HUMANIST: . . . is going to work with the masses . . . tomorrow's doctors, lawyers, teachers, salesmen, businessmen, and politicians. . . . they are hyperactive.
2. CONCEPTUAL: . . . is more into projects than people . . . tomorrow's engineers, architects, designers, astronauts, pilots . . . they're not clumsy in their body, and they're often very athletic as children. They have control issues, and the person they try to control most is their mother, if they're boys. Girls try to control their fathers.
3. ARTIST: . . . much more sensitive and oftentimes smaller in size, though not always. . . . They are creative, and will be tomorrow's teachers and artists.
4. INTERDIMENSIONAL: . . . *they are larger than all other Indigos* [emphasis added], and at one or two years of age, you can't tell them anything. They say "I know that. I can do it. Leave me alone." They . . . will bring new philosophies and new religions into the world. They can be bullies because they're so much bigger.

The minute they start talking, you talk openly to them. Get them to talk things through . . . [even] when they're babies. You chat with them—you talk things through with them. . . . all they ask for is to be respected as children and to be treated as human beings. . . .

Travis is an Indigo artist. He has been gifted with musical talent. At age four, he played his first public concert on the mandolin. He developed a young Indigo band at around age five, and after winning contests nationally at nine, they recorded their first CD. At 14, he had a top-ten hit off his solo album. All the songs were written, arranged, and played by him. He is considered to be, according to the music critic of the *Chicago Tribune*, the Mozart of the mandolin. . . .

Do you feel that your child really is smarter than you were, or than other children you have raised in the past? Perhaps the "smarts" are being diagnosed as a problem, when the reality is that they should be seen as an asset. . . .

In my (the author's) weekend courses, there are often some mothers, also grandmothers, of Indigo children, who "know" these young ones are surfacing. We exchange ideas and are able to help each other deal with these very special children with love and a greater understanding.

Doreen Virtue, Ph.D. says: "We know that Indigo Children are born wearing their God-given gifts on their sleeves. Many of them are natural-born philosophers who think about the meaning of life and how to save the planet. They are inherently gifted scientists, inventors and artists. . . .

Many gifted children are mistakenly thought to be "learning disabled" . . . [and many] are being destroyed in the public educational system . . . falsely labeled with ADHD. And many parents are unaware that their child could be potentially gifted."

. . . [Here are a few] characteristics to help identify whether your child is gifted [there are more in the book on page 24]:

- Has high sensitivity.
- Has excessive amounts of energy.
- Bores easily—may appear to have a short attention span. . . .

In summary, here are two main ways to identify Indigo Children:

1. If the Indigo Child has been identified as a "problem," testing is essential. While not all Indigo Children will test in the "gifted" range across the board, most if not all will exhibit at least one area . . . in the very superior range. School-based performance, more often than not, will be at least in the average range.
2. If a child is believed to be ADHD, chances are they are Indigo! Look for a range of "disruptive" behaviours that others mistake for ADHD. Indigo Children will be labeled as hyperactive troublemakers who won't "listen," since the old ways, such as direct requests, will not work.

[Also, we are reminded that Ritalin has enormous side effects. These are documented in the book by these two wonderful people helping many.]

Working with Indigo Children is akin to working on ourselves. The lessons they teach are obvious! . . .

Parents, we all want to tell you that THERE IS HOPE! [emphasis added].[14]

There are at present school systems that really do work with Indigo children. One is the Waldorf system, and the other is the Montessori schools. Here are some sources of information regarding alternate "Indigo"-type schools all over the world:

www.Indigochild.com
www.ch.steiner.school.nz/directories/frames/waldir.html

American Montessori Society, 150 Fifth Avenue, New York, NY 10011, Tel (212) 924-3209.

Association of Waldorf Schools of North America, 3911 Bannister Road, Fair Oaks, CA 95628, Tel (916) 961-0927.

Help your Children through Colour

Ask a child what colours they like and this will tell you much about their personality and interests. Up to the age of seven, a child is very close to the spirit world; they are spontaneous and natural and it is a time when those who look after them can influence them the most. (See my book *The Healing Tones of Crystal Bowls*, hereinafter referred to as *THTCB*, page 73.)

A knowledge of the psychological power of colour is important. A mother we know painted her son's room all Blue as he was so overactive, but it had the effect of making him feel depressed and yet still overactive. We recommend a pale Blue with Peach—the Peach will still have a calming effect without being depressive.

Pink is a lovely colour, and not just for children. Pink is the colour of God's Love and we all need to be loved. You can use Pink by imagining Pink petals falling over the child—it will have a calming effect and more open response.

It is good to see Orange being used more today. This colour has Red and Yellow in it, the Love and Wisdom of God, and brings joy and upliftment. This colour also strengthens the etheric body. Orange will help autistic children. At a daycare centre in London, England for handicapped children, they introduced Orange cushions and found a greater response from the children. They said it was very noticeable. An osteopath painted her small reception area Orange and the patients love it; they feel warm and secure.

A quiet child would respond to sunshine Yellow with soft Green and Pink/Orange for a bedroom. The room in which they sleep is most important for children, but also for all of us—we need to be able to relax and feel refreshed when we wake up. Yellow here would bring a child out of him/herself while the Pink/Orange gives one a sense of warmth and security, and the Green brings balance.

For children who do not sleep well, try a Pink light—this will soothe and relax. For the child who lacks concentration and is at school, use a Yellow light.

The Zen of Parenting

The following fourteen pointers on parenting are found in Janet Hobbs' book *The Gateway*:

1. It is possible to reach the Divine through the simple deep love of one's children.
2. You will get back from your children what you deeply give them. You know what you deeply give them, so don't be surprised at what you deeply get back.
3. The sign of a good parent is the ability to forgive self, because we are all bumbling human beings. The more we truly forgive our mistakes, the more we can allow our children to be human too. Whatever we can forgive in ourselves, our children can forgive in themselves too.
4. The doors of Heaven open when a child is born and that's why there's often such joy. Keep reminding yourself of that feeling when your child was born and seemed a gift of God. Be faithful to that feeling of love for your child.
5. Let your children tell you the truth of how they feel, even if it is upsetting. They often put words to what you deeply feel anyway.
6. Sometimes it is easier to love one's children than oneself. Let the children bring out the best in you and then give that best back to yourself.
7. The more you love and respect yourself, the more you will instinctively love and respect your children. If you are harsh with yourself, you will be harsh with your children. Either overtly or covertly.
8. When you are angry with your children, look at their freckles, their little lips, their cheeks, their vulnerable necks, their soft bodies, anything to help you reach past that moment into your immense love for them. The more you can remember your love for them, even when you are angry with them, the more fulfilled you will be, both as a parent and a human being.
9. The feeling of success comes most powerfully from the sense it is safe to be fully oneself. Someone putting on an act could be Prime Minister and still deeply feel like a failure. The most valuable achievements come from people whose work is an outpouring of themselves, whose work just couldn't be done by anyone else. These people leave their mark. We could raise a generation of children like that and the world would be transformed forever.
10. Whatever emotional garbage you deal with, won't be passed on. The most loving thing a parent can do is to confront his or her own demons rather than projecting them out and passing them on. It becomes joyful to do this for one's children.
11. Allow life's hardships to bring out the best in you.
12. Children are Divine Beings. So are you. But they are fresher from Heaven and deserve respect for that.
13. If you let yourself learn from your children, your old age will be happier and probably more healthy.
14. You were once a child.[15]

Meditation: The Simple Daisy

Conjure in your imagination a beautiful white Daisy. . . .
See it against a blue background . . . perhaps the sky. . . .
Perhaps you are holding it or maybe you have it
in a special vase near the window. . . .

Hold still . . . see its perfection . . . its simplicity . . . its naturalness and beauty . . .
white petals, slightly ridged for strength, palest pink-tipped . . .
its yellow centre—such a simple design for its beauty
to shine through to your heart. . . .

See its heart centre as full of promise . . . golden . . . pure . . .
sending a message of simple clarity to all: BE—simply BE YOURSELF . . .
have a simplicity . . . this pure quality of Light . . . strength . . . all-knowing. . . .
Be a shining Radiance in the blue for everyone,
so that each can feel your beauty and have the same childlike faith
of the simple Daisy in Nature. . . .

Hold it to your heart. . . .

Chapter Five

Teenage

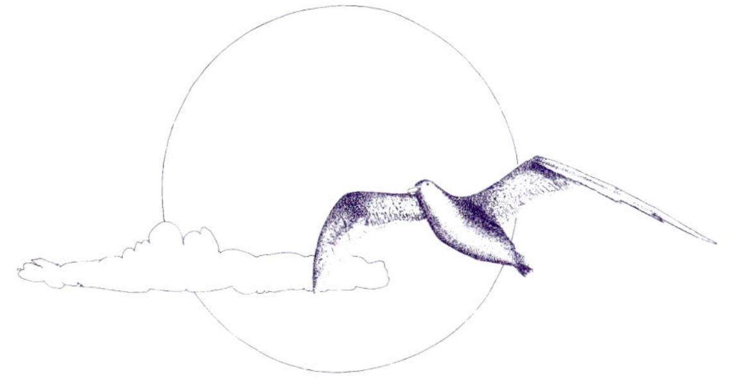

In his *Theory of Colour*, Goethe, one of the world's most profound mystics, gave to man a textbook on this subject which will be a classic in the incoming New Age. To him color was the Voice of God speaking through Nature. Blackness was not merely absence of light; it was the background of the cosmos, a field of intense activity for Beings of a vastly higher order than our humanity. . . .

White light is weighted with the power to bless. It is God speaking through the manifold activities of creation. When man learns to tune in and freely utilize this transcendent force, many limitations and disabilities of this three-dimensional world will be conquered or eliminated. Then will man assume his rightful place as . . . a true son of God in full possession of his divine heritage.[16]

—Corinne Heline

Within the Light

In Nature, while you were walking, have you ever felt that if you could just keep going towards the sunlight you could be completely immersed within this beautiful, golden, warm light of the Sun? That you would be so enfolded, filled, protected, warmed through, loved, that nothing else would matter?

This is a Truth you have felt, many of you, that within that Light is Everything—all the sustenance needed for daily life, all the protection, the encouragement, the caring, the Love you crave. It is all there within the energies of Light, and to reach forth to it, to go into it in our mind's eye is the very first step to be within that Light. To ASK for it to encompass you, then to feel it penetrating with the breath into every cell—to be able to move and have your very Being within the Light of the Sun—that is LIFE, that is all that is necessary.

Daily, then, let us aim towards the Light. Let us consciously breathe it into our very Being, into each centre, especially the heart. Let us be aware of this precious Light, and even if the Sun is not shining, know beyond doubt that it is still there, hiding behind the clouds and its energies are still permeating the Light you can see.

Let us make it a daily practice to absorb the Sunlight, to uplift ourselves with it in the very air we breathe, and let us feel the "sparkling" within, knowing that we are truly nourished by the powerful God-given rays of the spiritual Sunlight.

THE TEENAGER

The most exciting time of life is in youth when we are "racing" to grow up and become a person in our own right, to be able to see clearly our Pathway. Meantime, we are beset by all sorts of desires, some which engulf us, yet our feet are firmly on the ground and we pursue our way each day wondering how to cope, joining with our very special friend for a closeness if only in understanding, feeling that no one else does, even—or least of all—our parent(s).

Have you ever thought of sending a prayerful thought to your Guardian Angel? For we all have one, who walks every step of the way with us, listening, loving, encouraging, if we will only listen. In this way we can feel we are not alone, we can get simple answers to simple questions daily.

If we make a "tryst" (a meeting in thought) daily—"Here I am, dear Angel of mine. Please give me a thought to guide me on my way this day"—and take life one day at a time, then the Path is made easier. We can begin to see clearly where we are, why we are on this certain pathway, why we are with certain people, and what we can learn from each as we go along.

Take heart, young friends. There is always one to whom we can talk, release our pent-up feelings, and even if not possible every day for whatever reason, then write in your special journal all your thoughts and feelings for that day. Written, things seem to simplify or evaporate, melt away in the sunshine, and we can feel easier, more buoyant and able to cope once more.

Try it for yourself. Give a special name to your Angel and commune daily. When you wake up in the morning, take a special moment to say, "Hello, dear Angel, it's a brand new day. Please help me to try again. Much love. . . ." And again before you go to sleep, send love and thanks for help during the day.

In this special way, you can feel cared for, loved, and you can give your love. Then can your days feel more balanced and beautiful. Know that you are much loved always. . . .

COLOUR IS SO IMPORTANT

If teenagers could only wear brighter colours, it would be so uplifting for them as individuals—also for their peers, because we are compelled to look at colours others use for their dress!

Read for yourself the effects of each lovely colour in this book. It will help make the daily decision of what to wear—provided of course, your wardrobe is not too confined to one colour, such as Black.

BLACK

There are some teenagers who crave Black in their room, where they also sleep. They want to discard all that went before, to hide themselves in the darkness of Black. Maybe they are on drugs, something is bothering them, there is despair; it may be the Dark Night of the Soul, and help is desperately needed.

If you are a teenager and cannot talk to your parents openly, then talk to a friend to "unload." Remember, wearing Turquoise helps to alleviate stress—it is for clarity—and warm lighting has a calming influence.

Many people ask me about Black. I was once especially invited to a birthday party to talk about Black. Everyone had been asked to wear a colour (other than Black), and some women who had complete wardrobes of Black had to go out and buy a new colourful dress!

I told them:

1. Black hides the dirt;
2. It is drab—it has no colour;
3. It is depressing to gaze upon;
4. It is a myth that it makes one look slim;
5. People who wear black today are like sheep gathering together to follow fashion designers—perhaps the Chinese designers, who claim that Black means wealth. The Japanese designers all seem to wear Black—it is their uniform!
6. Black is seen everywhere we go these days. What are these people mourning? Do they realize how very depressing it is to see people all in Black from head to toe—in restaurants, at the ballet, the opera, the symphony, in the supermarket and on the street. And on rainy days, Black seems to add to the dreariness. What a chance we each have to uplift people with lovely colours we can wear! But let us absorb guidance from the following (sometimes contradictory) research on the colour Black:

- The Light contained within Black is turned in upon itself, and the lowest aspects of one's nature take over. People have never really understood Black—some associate it with mystery and the unknown.
- Black is also said to be one of the Highest colours, and it is worn by advanced Initiates of secret mystery schools, symbolizing mastery over the physical plane!

Let's see what Tony Cooper, practitioner member of the International Association of Colour (IAC), has to offer about Black:

> Black relates to our collective consciousness (mainly unconsciousness), to Saturn . . . and often to bad past life experiences felt through the Astral body. Black serves those who pretend to be what they are not. Everyone knows that black clothing is thought to be slimming, and a black cocktail dress is seen by many to be sophisticated and fashionable, or even a sexy little black number. But in reality what it is saying is I need to absorb all frequencies, I need special attention.
>
> Black implies weight and solidity; darkness implies infinite space. Any black material, even velvet, reflects at least 3% of incident light. Black is inherently ominous as it is the unknown, the mysterious. When black is worn with other

colours, the other colour appears more than twice as important. People who always wear black clothing are generally turning away from what society has to offer—in part because of dissatisfaction or disenchantment with the system. Black then becomes a symbol of the negation of society's values. This negative response then needs a positive replacement of the rejected values. Such people are looking for a leader or role model to present them with more acceptable beliefs.

People who are lost identify with black standing for the lack of light or direction. One needs to let go of earthly fears, for only fear separates us from our divine love spark, the God within. Know that love is our birthright. Live in that Love. Black as an adjective—blackmail, black market, blackball, black-listed, black sheep, black looks, black plague, black Monday, black Maria, black magic.

Black absorbs all colour, but is not a colour within itself. Black does not contain all colours, and it cannot be made through the mixing of all pigments. It links to secrets, negation and renunciation the absolute boundary beyond which life ceases, surrender, a hiding place, "I need special attention," withdrawing from the unhappy experiences within society, hiding one's light (potential) under a bushel, finding present circumstances disagreeable or over demanding.

Finally, I would like to remind everyone that black, like all true colours, is just what it is—neither negative or positive. It is how we perceive and use colours that gives them their special qualities.[17]

In summary, probably Black is the most misunderstood colour—but above all, decide for yourself, dear Reader, what you feel about it all, and take what you need for the moment as your Truth. Leave the rest until there is greater Light on the subject!

Coloured materials only allow like colours to pass through.

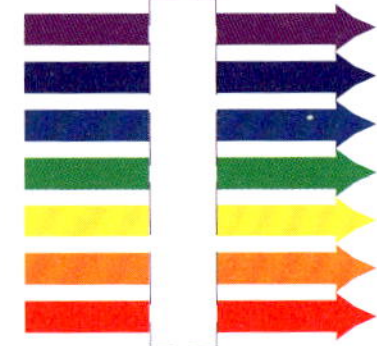

White materials allow all colours to pass through.

Black materials absorb all light and colour.

Figure 3.

White and Black

For centuries, for aeons of time, these two, White and Black, have represented opposing strengths: White stands for purity of purpose and Black for obscurity, a protective power of the hidden. They also stand for the opposition of Yin and Yang, feminine and masculine, delicacy and strength.

While White can be said to contain all colours of the spectrum, Black does contain some, but absorbs them all into the materials worn, so that no Light can enter the body. From a healthy point of view, then, to wear White allows all colours into the body, through the pores of the skin.

So climb back on the ladder of Light. Choose brighter colours for dress; be aware of the true meanings of glorious colour; become alive with Light— breathe it in from the Sun. Allow your beautiful body to sparkle, all the cells to dance, and then everything falls simply into its rightful place. You are whole, you ARE the Light, and you are loved.

Schoolchildren's uniforms, if worn, need to be analyzed. If Grey, children tend to be judgemental, uncooperative, have poor results in exams. If Blue, students are more cooperative, helpful and exam results are good.

Those who let you down . . . bless them—perhaps they are giving you a gift.

White

The White Ray is a great purifier and cleanser. It shines as all colours together in a powerful way. When we add White to other colours, we purify them into what are called "tints." The colours become softer and take on more enlightened qualities.

White alone shows us our imperfections. Too much Whiteness can make us feel uncomfortable and we need other colours to soften our environment and remind us that we are human—perfect in our imperfection.

We need only look at a waterfall or a snow-capped mountaintop to recognize the pure cleansing quality and the power of White.

Wearing White often, one may have the tendency to become a perfectionist; friends may find it difficult to relate to you. Wear some other colour WITH IT to keep you "earthed" and in touch with reality. White does radiate, shine and eliminate all that is dark.

People who are channels for healing often instruct clients to surround themselves with the White Light as a form of protection.

This is the Light which casts out darkness, preventing it from penetrating. There is great power in working with the Light in this way.

Waves of Light

In our hearts we are aware of rhythms—the beat, the coming and receding, the warmth and the chill, the rising of expectation in feelings and lowering of the warmth by fear. These feelings are all caused by waves. We are enveloped by waveforms—they come and go, they march through us in a flowing movement, they touch us within and without. We breathe them in. We also breathe them outwards to others.

These waves are life-giving. They have within them the quality of Life—of pranic force. They enhance our feelings—the life force within. The more receptive we are to them, the more life-force is there for the taking.

But what if we are not receptive to them, you might well ask. What if we are "dead" to these forces? Completely unaware consciously? They work on the unseen level, they ebb and flow through us, like an invisible web. We can receive these waves through our subtle bodies first, then our denser body through the skin and through the breath. They are full of myriads of colours, each colour touching us where it may, and if we could be aware, we would see that our aura is filled with these waveforms. They sparkle and dance all around us—sometimes even on "grey" days we can feel very uplifted and wonder why.

It is almost as if there are Angels dancing around us, enveloping us with their chiffon-like energies—encouraging, uplifting our spirits.

From now on, then, let us be more awake and aware of them—ask ourselves, what do I need today? And take deep breaths of thankfulness, for is it not true that everything we need for our true sustenance is at hand? We are so cared for, let us not doubt it any more, instead simply find that moment of stillness within and receive whatever is there for our sustenance. Then can we take the next step onwards, in the wave of Light enfolding us with Love.

Meditation: Into the Light

As you breathe yourself quietly into a meditative calm . . .
begin slowly to visualize a pinpoint of bright Light in the distance—
it sends forth a beam to you.

Concentrate on the point of Light calmly and watch it. . . .
Gradually the Light begins to grow, to send forth more rays . . .
to move slowly towards you. . . .

Allow this pure white Light to approach you . . . to enter your Heart centre. . . .
Allow it to penetrate . . . to grow and expand. . . .
Allow what is not Light to be shed. . . .

Let this beautiful Light permeate your whole Being . . .
then radiate from you—wider and wider—enfolding all near you . . .
expanding out into the Universe . . .

until ALL IS LIGHT . . . and you are ONE with it. . . .

Chapter Six

Your Beautiful Auric Field

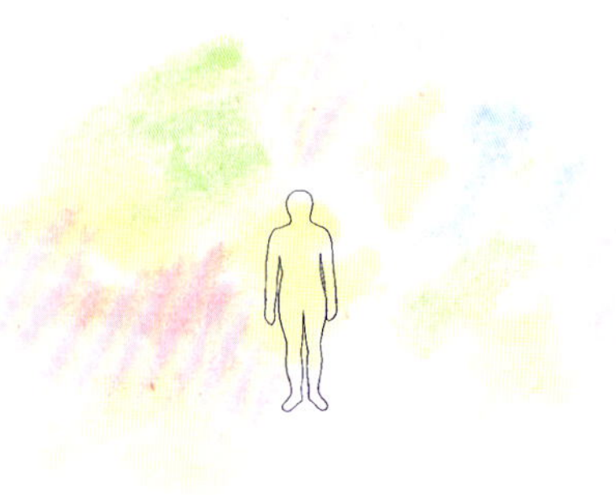

> Be thankful. Whatever may be your circumstances at the present time, be thankful, for they are the soil about your roots, from which the blossoms will spring forth in due season.
>
> —White Eagle

Your Beautiful Aura

Have you ever contemplated how beautiful is your aura, if only you could see it? Have you seen yourself in a mirror, closed your eyes for a second, then opened them again to take a peek at the colours around your head and shoulders? Or, can you see them around others if against a white background? Realize that because of all our experiences (which in themselves create growth), the colours may even become paler and more beautiful. Sometimes others may catch a glimpse of your colours too, and remember, even if we cannot see the colours, they do "touch" us when we are within a person's auric field. We can feel "touched," we can feel uplifted or we can feel uneasy, edgy, not wanting to be there.

Think about this when next you are near/with someone. What do you feel? Are you truly delighted to be with them? Look into their eyes—how does it feel? Are they open with you (honest)—or cagey, looking away all the time? That could also be shyness, but what does your heart tell you? Lovely people tend to exude a wonderfully peaceful feeling, and often a gentle perfume to go with it.

Sound too can enhance this feeling, make it grow larger, all-encompassing, or shrink and stay close to protect one. Have a thought about this and next time you are with someone or meet them, ask yourself, "How do I feel? Do I want to be near or far away from this person? Can I relate, or do they need help with energy through (not from) me?"

If you are drawn to read this, you are a Light Being. People feel your emanation whether you are conscious of it or not, and they will leave you feeling really great. We can give great thanks and Love for this opportunity, and go our way rejoicing in the Light.

The Auric Field

Our sun generates electromagnetic waves in all directions. The different forms of energy (i.e., all the rays) can be identified by their wavelength and frequency. Wavelength and frequency are related: the longer the wavelength, the lower the frequency (Red); and the shorter the wavelength, the higher the frequency (Violet). A light containing the full visual spectrum will appear White to the eye.

The Kabbalarians, a Jewish mystical theosophy group that began about 538 BC, refer to energy around the body as Astral Light.

Reasons for the above:

1. The sun's product is energy in wavelengths;
2. When all wavelengths are present, then all three colour producers in the cones of the eye are activated, so we see white light;
3. When a part of the visual spectrum is absorbed by any substance, i.e., the ground, trees, fabrics, etc., the remainder received by the eye from the visual spectrum is now incomplete and the difference is recorded in colour by the cones.

Paracelsus (Swiss physician and alchemist—1493–1541) studied and worked with energy fields which he called "Illiaster," composed of both vital force and vital matter.

Remember, there is no colour *out there*, only energy in the form of waves, rays.

HEALING—
EMOTIONALLY, MENTALLY AND PHYSICALLY

Illness is a sign that the Body and Mind are not being used properly. Attention is not being paid to the lesson needed to be learned—therefore there is DISHARMONY.

"Our sickness and our pain are gifts from the Universe . . . prodding us to come into balance. . . . This is part of the perfection of the moment."

—Laurene Johnson,
channel for healing

The word AURA comes from the Greek word meaning "breeze." The Aura is made up of attractions—we attract what we deserve. When we radiate, we attract vibrations from around us, from the Universe—plants, people, etc.

As we emanate vibrations from the chakras in bursts or steady streams, they flee the auric field, coloured by our feelings. When we need a colour, we are subconsciously drawn to it in our clothes, food and other people who are that colour.

Our overall colour changes with moods, places we are in and the people we are with, and when the auras blend, there is an interchange. We are surrounded by ether. We are flowing and floating in it. It is pliable and each time we think, it forms into shapes. These thought forms float around us and crowd the invisible world.

It is often uncomfortable to be with one whose mind is churning—negative thoughts are shapes that flow into the aura and interrupt the positive vibes from the cosmos. Negative thoughts attract negative vibes from the Earth and those around us, so the aura becomes a storehouse for negatives. If we worry, these vibes are left to hover in the house. Eventually the physical body is affected.

The Aura constantly reflects changes within physical, emotional and mental makeup. These changes show in the aura as colour variations, so our aura is always changing. Perhaps when it is really strong it does not need to change, but when you walk into a room IT changes, not you. A more spiritual person will restore her partner's aura, not impose her own.

Understand that you will never be really well until you are capable of forming a pure and powerful aura with all the colours of the spectrum around you. . . .

With white light you can have the all-powerfulness of violet, the peace and truth of blue, the riches and eternal youth of green, the wisdom and knowledge of yellow, the health, vigour and vitality of orange and the force, activity and dynamism of red. . . .

But above all, see this light as white. If you succeed in concentrating on light, feeling it as an ocean which vibrates, throbs and quivers, where all is peace, happiness and joy, you will also begin to feel that this light is perfume, and music, that cosmic music of the spheres, the song of all that exists in the universe.[18]

—Omraam Mikhaël Aïvanhov

TOOLS NEEDED TO MOVE CLOSER TO HEALING OURSELVES

- Breathing
- Relaxation
- Meditation
- Ability to feed from the colours of Nature, plants, flowers.

Awareness will change the body. It will become lighter and younger, as it opens to God Consciousness. Our aura then becomes a place of beauty and people will be drawn to us.

The Simplicity of Life

From the smallest dancing light on the waves of the ocean to the tallest tree in the forest, know that the intelligence of the great Cosmic Spirit is there within—and so it is with each human being who walks the beloved Mother Earth Planet.

We are covered from head to toe with light of many hues and shades. This Light also dances, unless we are very sick, and as it does so, it bounces off onto others whom we meet in our daily lives, touching them where necessary, whether slightly or deeply, according to the need. So in the absolute wonder of creation, there is rhyme and reason for everything.

Where one is sad, a Joy can be conveyed from another to touch that one, and for this reason, we must be open, alert and very watchful, mindful each day, aware of each and every thought, be absolutely sure it is a good thought, so it will be carried on the tide of our energies to whomever is in need. Thoughts do carry and they are cumulative—like adds to like—so from now on, this awareness must be sought and acted upon. When we clear and clean our own thoughts, then the thoughts abounding on this Planet are also cleaned and cleared.

Trust then, from this moment on, that you KNOW you are cared for richly, you ARE in good hands, and that the Higher Beings ARE watchful and care for each of us in whatever way is needed. What each needs to do is to be mindful, to be more accessible, to listen in each day, to trust what we hear or what comes into our mind, then act upon it—to work towards loving more, having more compassion and above all understanding. For "no man is an island," therefore it is true that we are part of each other. Our energies dance, so let that dance be full of Joy and the Light of the Sun!

Mother Teresa of Calcutta had Violet in her aura—wherever she walked she cleared everyone. Violet also helps clear unpleasant smells.

Meditation: The Calm Sea

Out of the mists over the sea there comes a shining Star . . .
small at first, just a beam of light . . . but as you watch it,
it becomes gradually stronger . . . more beams . . .
and begins to move towards you.
The mists begin to clear because of the light. . . .

The sea becomes so calm . . . and the light of the Star is reflected across the water.
Silence pervades all. . . . The silvery quality of the beams dances upon the sea . . .
and you feel in your heart centre such a stirring . . . yearning . . .
and from the depths there comes a feeling of Joy . . .
Joy for the gift of Life . . . Joy for the pureness of Nature . . .
Joy for the new day yet untouched . . .
Joy for the next challenge and the strength to go with it. . . .

And as you watch the beauty of this scene . . . you suddenly become the Star . . .
YOU are radiating Joy and strength and Love all around you . . . everywhere . . .
until all is One in the Radiance.

Chapter Seven

The Philosophy of the Chakras

Music gives a soul to the universe, wings to the mind, flight to the imagination, a charm to sadness, gaiety and life to everything. It is the essence of order and leads to all that is just and good and beautiful. . . . It is moral law.

—Plato

THE CHAKRAS

Firstly, we have to understand the principles of the structures, that they are vehicles in the etheric body for the passage of energy, so vibrant and scintillating . . . that the whole Being is energized with colour, and the particular colour has its own frequency, its own note or sound, and this sound can be in tune with the Universe, the great Cosmic Being, an harmonic note which is all-encompassing, is healing, is JOY, is Light, is Love.

This means then, that if we can maintain this certain frequency which is our very own keynote, then we are in harmony with absolutely everything, with the Great Universal Mind, with the Luminous Beings who accompany us on our Earthly Journey, and then here on Earth with all those with whom we walk, we touch and serve.

This then is the Philosophy of the Chakras: to be in harmony with all things great and small, to see goodness in everything and everyone. Then will the energy flow through us so evenly, so colourfully and so in harmony, each Centre one with the other, that eventually all Centres on this Planet shall join together with those beautiful ones on other Planets, thus creating the Harmony of the Spheres.

The feet and hands are the first chakras to open and, very importantly, they collect gravity energy from the Earth and root us, giving solidity and balance. That we must be grounded is so very important. Walking is a grounding experience and we must keep connected to the Earth to allow releasing of negative energies and let in earth energy. We can keep our energy field filled all the time using deep breathing, Yoga, Tai Chi, Qi Qong, Reiki and by toning every morning.

INFORMATION ON SUBTLE BODIES AND CHAKRAS IS MOST IMPORTANT BECAUSE OF THE GREAT ROLE THEY PLAY IN HEALTH, DIS-EASE AND SPIRITUAL GROWTH.

One of the important steps in modern medicine will be to understand the role of the subtle bodies and chakras in health and dis-ease. Dis-eases usually settle in the subtle bodies, especially the etheric body, before affecting the physical body. If imbalance can be treated in the subtle bodies and the aura before it enters the physical body, resolving the problem is much easier.

When people become more mature and their vitality lessens, imbalances in their subtle bodies and the aura tend to move into the physical body more easily.

Sometimes what happens when people have various adverse symptoms and a battery of lab tests proves inconclusive is that the imbalance is moving through the etheric body into the physical body. (Rudolf Steiner and Alice Bailey in many books provide real detail on the proper role of these bodies in our daily lives).

The physical and etheric bodies are so intertwined that there is no focal point. The Emotional Body is usually connected to the Physical through the stomach, Astral Body through the kidneys, Mental Body through the left brain, Soul Body through the pineal gland, Spiritual Body through the pituitary gland.

Rudolf Steiner, Madame Blavatsky and Ronald Beesley all suggest that the great Initiates have always expressed themselves with caution and have given only hints. Above all, they leave some work for the human being to do!

The purpose of this special work is to stir up the latent knowledge, the pre-knowledge that is already in each consciousness, to release that pre-knowledge and bring it into action now. . . .

We are children of the Light, we are composed and fed from it . . . we are but on the fringe of bringing . . . beauty, of texture, colour, form, shades of such glory and wonder . . . into a colour-hungry, Soul-starved, frightened, sick world.[19]

—Ronald P. Beesley

Newly Awakened Chakras

In my first book, *The Healing Tones of Crystal Bowls*, I wrote about nine chakras. I included two newly awakened ones (Thymus and Zeal Point). Since then, very enlightened souls have talked of more newly awakened centres and written about them. Sheldan Nidle presents a chakra system of thirteen centres in his book *Your First Contact*. He includes the Thymus (#6) and the Zeal Point (#8) at the nape of the neck (calling it the "Well of Dreams"); he also adds a new one for the Diaphragm, and two above the Crown Centre (see Figure 4).

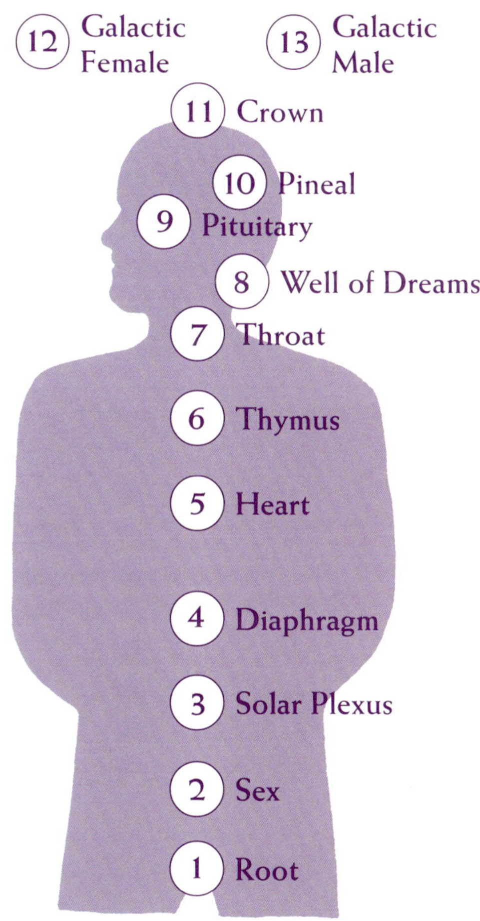

Figure 4. *The New Chakra System.*[20]

The Truth about Chakras

Anodea Judith writes clearly and succinctly on this subject:

> Chakras are sometimes referred to as lotuses, for they open and close like a flower, and in the Tantric system they are shown with a varying number of petals. The petals, ranging from four at the base chakra to 1,000 or more at the crown, express their vibratory rate.
>
> When a chakra is closed, the life force energy cannot travel through that part of the body, and one might say that the programming in that chakra is locked in a restrictive pattern. If this is the case, we feel a lack in our lives in its related area (such as the ability to communicate, Chakra #6 [Throat]), and our physical health in the chakra's related functions may also be affected (sore throat, tight neck).
>
> A chakra can also be "overblown" if it is out of balance with the other chakras in the system. In this case, that particular chakra uses so much of the body's energy and the mind's attention that other areas become deficient. An overblown third chakra [Solar Plexus] causes an attachment to holding power over others, hindering the ability to find the love and balance associated with the heart chakra directly above.
>
> With attention and understanding, we can control and influence our chakras. They can be developed like muscles, programmed like a computer, nurtured like a seed, or closed like a book. Development of the chakras occurs through understanding the system as a whole and then working directly on specific areas. Techniques may include physical exercises, processing of old traumas through therapy, visualization and meditation, chanting of mantras, working with their elements, herbs or gemstones, and personal ritual, as well as the general lessons that are brought to us through our daily lives.
>
> The body is a vehicle of consciousness. The chakras can be seen as the wheels of life that carry this vehicle through its evolutionary journey towards enlightenment. Within us these wheels are like gears, each one appropriate for different activities or stages of life. As we open our chakras, we become more conscious and more fully alive. Our journey becomes smoother, more productive, yet more exciting as we become more fully who we are.[21]

The Rainbow Dance

You have your eyes, you have your ears: look with your eyes on the things of Nature, hear with your ears what goes on in Nature; the spiritual reveals itself through colour and through tone, and as you look and listen, you cannot help feeling how it reveals itself in these.

—Rudolf Steiner

There are five minor chakras in the hands and feet—and remember the Zeal Point chakra mentioned above, situated in the hollow of the neck at the base of the head where the brain stem (the medulla oblongata) meets the spine.

There is now a version available—by Jan Linch—which also numbers thirteen chakras, but a little differently (see Figure 5: Geometrical Chakra System). Here we assume that the Zeal Point chakra is accommodated by #9, the Tongue, associated with the colour Sapphire. Linch's chakra system includes centres for the Pancreas, Lower Heart, Tongue, Cerebellum and Astral (just above the Crown). Each one of these embraces lovely added colours.

What we must do now for ourselves is to decide intuitively which method to accept, or whether we shall combine them—and if we have any intuitions about the right colour for ourselves or a patient at the moment, then to use that one. There is no hard-and-fast rule, and Linch suggests the Thirteenth Chakra is the beginning of a whole new octave.

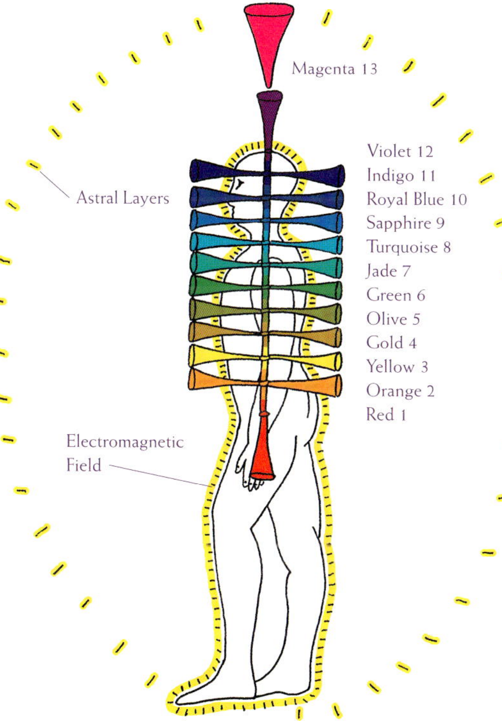

Figure 5. *Geometrical Chakra System.*[22]

Remember, these are major chakras—we have at least 350 minor chakras too, including the ones in the hands and feet.

Never Underestimate the Healing Power of Sound

Jan Linch writes on the connection between sound and the chakras:

> Sound also plays an important part in the vibration of energy. Each chakra will resonate to a sound frequency and, therefore, chakras can also be put back into balance with music. . . .
>
> If you play the notes individually on a keyboard it can help . . . You may find that your favourite sound frequencies resonate with your favourite colours.
>
> Your chakras enjoy sound frequencies and can appear to be stimulated or relaxed by certain tones. How often are you moved only by certain pieces of music or certain chords, only at certain times of your life? When you enjoy listening to music you send out unconditional love which raises the vibration within your aura. Even heavy metal bands can have this same effect if this is the music you really enjoy. Once unconditional love is sent out into your aura the chakras respond by accepting and resonating with the healing vibration of sound. How many of you actually relieve the stress of a hard day by putting on your favourite piece of music? Never underestimate the healing power of sound. You may be aware on a conscious level of music therapeutically making you feel better but on a subconscious level your chakras are being put back into balance and are, therefore, being healed![23]

The Gossamer-like Patterns of Life

Have you ever wandered down Memory Lane—as far back as you can in this present life—and wondered about the beautiful pattern beneath all that has occurred in your life? How one step builds upon the next—or rather, the one before? How, if you had not been "prompted" to take a certain step—then the next one would never have been built? And how it seemed you were miraculously led to the next step—perhaps an incredible meeting with someone special and you learnt from that?

Have you ever wondered how your life in retrospect could be so "neat" in helping you learn such marvellous teachings—all in simple relationships and happenings—and how, if you had not done such and such a thing, you would not have been introduced to a "new" person or new ways of thinking?

These gossamer-like threads, often seemingly fragile, but truly really strong, are the very threads of our existence here. They lead us onwards to the next connecting thread, and in time we are woven inextricably with other beautiful souls; we connect at the appropriate time, and we breathe a great sigh—for are we not so cared for? Perhaps most only recognize this in dreamy retrospect; nevertheless, this heavenly thread is pure gold and silver. If only we can be watchful, mindful, then we shall be shown in our daily vision the Golden Way. We encourage you all to stop, look and listen and follow your heart wherever it leads onto your Golden Pathway of Light.

Meditation: Journey to a Rainbow

There is a soft mauve Light around you. . . .
It has a warmth and a quality of calm—tranquillity. . . .
It has a perfect stillness . . . and out of its beauty
you feel a great sense of wonderment . . . upliftment . . .
enfolded in this beautiful, harmonious colour. . . .

You begin to drift . . . lightly floating . . . as if in a cloud. . . .
You find yourself in your Body of Light moving along a RAINBOW . . .
within the pale mauve arch. . . .

There at the end is a great golden Light sparkling . . . the Great Sun! . . .
many-coloured beams within its rays. . . .
YOU become a Beam of its Light . . . bright . . . very beautiful . . .
joyously sending out your Love to whomever comes within its rays. . . .

You ARE Love. . . .

Chapter Eight

Sound

I envision the day in the not-so-distant future . . . when singing, toning, chanting and other forms of music echo through the corridors of every hospital unit.[24]

—Mitchell Gaynor, M.D.

Music Hath Charms

"Music hath charms . . ." is a very well-known saying—and what does it mean? Can it have any deep significance for our learning of the Spirit? How can we interpret it and relate it to our heart centre? What is it in music that charms us, touches us deeply, draws the strings into melodies which reach deeply into our hearts and memories?

All life is a song. We have many melodies that we use. Some are joyful, light-hearted like a touch of Spring, others more muted and well-balanced, quiet as a Summer's day without a breath of wind. Still more are muted, solemn, sad, touching us deeply where it hurts. And then there are louder chords banging away, stirring us to action, to march with the current mood, to be strident, to shout from the rooftops with gay abandon—all these in the course of our "musical" lives, tempering, uplifting, leading us on our Pathway, and always touching the strings of our heart, so that we move our Beingness to the melodies, the songs, and we are healed—made whole in every way—often unexpectedly.

It is said that the sound of the violin touches our heart. This is true. Its vibration touches our own heartstrings and plays a melody there, sometimes very sweet, sometimes melancholy, often joyously; it MOVES us to vibrate in tune or to action, re-action, or else it soothes and restores.

Listen then, whenever you can, to music and learn to listen from the heart, for deep within the heart is a place of recognition, of a knowing, and an openness if only we will allow. Open the doorway to your heart, leave the door ajar that other musical sounds may enter and touch the strings within, joyously—so much so, that you feel you must now open the door wide, be cleansed, refreshed and touched to share your Joy with those around you, and in this way more people will then feel able to open wide their doors to the deep mysteries within their hearts too!

SOUND

When we talk about Sound, we need to emphasize and be aware always that we are so very individual in every way. The Sound we need to hear for the moment may not be the choice of another, even though the ailment to be healed is the same. And so there follows a choice of many avenues or suggestions from which can be selected the Sound, melody or tone to use for your very own. Use your wonderful intuition for your needs at the moment!

So many people have expressed their diverse views, thoughts and intuitive impressions on this subject that I feel compelled to share what I consider to be the best of them with you in order that you may form your own opinion. One of the finest articles I have encountered is by Robert Willix, Jnr., M.D. and Ayurvedic healer, in one of his monthly newsletters. It is quoted at length.

> Scientists know that every atom vibrates, much like the strings of a guitar . . . and because we are a collection of atoms, we too vibrate . . . and also generate sound. Some is audible—I can hear your heartbeat through my stethoscope and can listen to gut sounds, lung sounds and the whooshing of blood through your arteries. . . . To a doctor a human being is a virtual symphony of sounds.
>
> The interesting thing is that science has determined that human beings when they're relaxed vibrate at exactly the same number of cycles per second as the Earth! (8 cps). One of the reasons that exercise is best done out of doors is that the vibrations of Nature—its music—can help restore your mind and body to health and harmony. It's no accident that we use the terms like "harmony", "in tune" and "tuned up" to refer to a positive state of mind and health. . . . But what most people don't realize is that the same sounds that make us cry or laugh or leap up to dance are resonating—vibrating—in every cell of our bodies as energy.
>
> It's the energy of sound that doctors use to blast kidney and gall bladder stones to smithereens. . . . Our tissues are mostly water, and sound is easily conducted through water to every corner of the body, generating sympathetic vibrations in our cells . . . which are the healing force that can turn unhealthy cells into healthy ones (and vice versa). This is not just theory. . . . Scientists have subjected living cells to the musical vibrations of a tuning fork. They found that different frequencies of the fork could alter the color and shape of the cells, e.g. a C note made them longer, an E note made them spherical, and an A note changed their color from red to pink. A similar test with cancer cells found that at higher frequencies they disintegrated as easily as kidney stones. . . . European studies have found that wounds bombarded with sound waves heal in two-thirds the normal time and that circulation, metabolism, nervous and endocrine systems are all improved by the application of sound energy [see *THTCB* page 111].
>
> I believe that someday all doctors will be able to diagnose disease in its earliest

stages by tuning in to the vibrations of the cells . . . that it's the changing motion in the cells that gives the first sign of imbalance, even before a disease like cancer or high blood pressure manifests itself. Your cells use their natural vibration as a way to communicate with one another and with the rest of the body. But when they are assaulted by a virus, bacteria . . . a chemical or constant stress, they begin to play "off-key". Not only does each of us have a healing Keynote, but each part of the body resonates at different frequencies . . . each organ has its own Keynote . . . [that] when heard, hummed, changed, or played, can restore that organ to its optimum vibration—literally put it back in tune.

Take a look at the following chart. It contains the specific organ Keynotes as identified by scientific research. . . . Say, for instance, you have heart disease. The Keynote for the heart is F. You can listen to a piece of music in the key of F, but for real therapeutic value, it would be better if you learned to chant, sing or hum the note, or play it on a musical instrument, preferably when you are in a meditative state. . . . You can use any sound—Ohmmm is probably the most often used, but you can use Aaahhh or Ohhhhh if it seems more pleasant to you.[25]

Keynote	Body Part
C	Lower back, sciatic nerve, hips, buttocks, lower bowel, legs, ankles, feet, prostate gland, bones, and haemoglobin: affects your emotions and self-esteem.
D	Lymphatic system, reproductive system, skin, kidneys and bladder: affects your energy level.
E	Nerves, muscles, spleen, solar plexus, kidneys, cells, liver, and intestines: affects the intellect.
F	Heart, lungs, upper body (including arms, shoulders & hands), glands, and immune system: is emotionally soothing.
G	Blood and circulation, throat and neck, spine, nervous system, metabolism, ears, and immune system: effective temperature control and stimulates an outgoing personality.
A	All the sensory organs, blood and muscles: pain and muscle control and can improve the intuition.
B	Blood and fluid electrolyte (mineral) balance, spleen: is helpful in meditation.[25]

KEYNOTE—OWN NOTE—SOUL NOTE

The Soul Note—is this your Own Note? Is it also your Keynote? Is there any difference between all three, and do we acquire them as we grow towards the Light, or do we arrive on Earth equipped with them? These are very good questions!

Firstly, it is an idea that our Keynote accords with the date of our birth and the planet or sign for that date—and you can see from the chart below which note "belongs" to each sign (according to Steven Halpern, Ph.D., musician/composer):

Aries	Middle C	Libra	F#
Taurus	C#	Scorpio	G
Gemini	D	Sagittarius	G#
Cancer	D#	Capricorn	A
Leo	E	Aquarius	A#
Virgo	F	Pisces	B

However, who made the original chart, and how long ago? Has anything changed since then? Take the following example:

I was born under the sign of Leo, so my Keynote could be E. This relates greatly to the Solar Plexus; however, I feel intuitively (for the moment, at any rate) that my note is G. Perhaps this would be my Own Note?

Let us delve a little further. I find that to listen to E is very, very balancing and uplifting for the emotional centre—whereas G greatly helps the throat centre. So perhaps G is needed more for the moment to balance the throat; then I can revert to the overall Keynote of E as soon as balance in the throat is experienced.

To find our Soul Note, we need to tune into ourselves, our Higher Self especially, and listen carefully, for the answer is there, and it may well be the complementary note to our Keynote. Would not this be a complete balance?

E = Leo Keynote = Yellow

B = Complementary Note to E or Soul note = Violet

G = Own Note = Blue

The Lapps believe that everything has a relationship to sound, a relationship that varies with age and changing conditions. So closely attuned are they to unspoiled nature that they even recognize the influence of the seasons of the year on the sound values of all natural phenomena. To Laplanders everything emits its individual Keynote. These are looked upon as so many fragments which, when unified, compose the symphony of nature. These people are fully aware that in this fact is a deep spiritual significance, and they desire to so think and live as to bring their being into harmony with nature's rhythms.[26]

—Corinne Heline

Once you have established your Keynote, add the body parts noted by Dr. Willix and the corresponding zodiac signs (as suggested above) of Steven Halpern to complete the picture. Of course, others have reached different conclusions, probably through different techniques and procedures—and researchers will continue to uncover more facts.

The combined sounds of everything on Earth compose a harmonious chord which is the Keynote of our Planet. It is the Note F, whose tone becomes visible as Green (Earth's basic colour note). Every object has its Keynote, giving concord between some and discord between others.

Frequent repetition of one's Keynote has a soothing, harmonious effect on the body. The high water content of the body's tissues (we are 70 percent water) helps to conduct sound, so repeating one's Keynote is like getting a deep massage at the molecular level.

If a person is able to discover his own key-note or chord and play it over gently to himself he will revive as if by magic. One's key-note can be ascertained by listening to some good orchestral music. When the note is played it will send a thrill right through its owner.[27]

—*Vera Stanley Alder*

Breath and Sound are of great importance in Voice Healing—they are totally interdependent. Voice Healing is particularly effective if practised in a group, for each resonates with each other, setting up an energy circuit.

Sound therapy boils down to discovering a "Universal" Keynote resonating to one's own individual Keynote. If ill-health amounts to being out-of-tune with one's own Keynote—then finding that note again produces perfect wholeness. Learning to identify, recognize and tune into the individual Keynote of one's Body as well as the notes of every Chakra and Organ, so subtle though they may be, is surely the medical science of the future.

—Sound Research Group, East Sussex, U.K.

. . . [I]f one may produce a sound which is on the same natural frequency as the soul vibration, there will be a transfer of soul energy. In fact, such a transfer does occur when individuals are able to produce their own personal sound. The pitch of your speaking voice is your natural sound frequency.[28]

—*William David*

YOUR OWN (PERSONAL) NOTE

This note you can use for balancing the body and for acquiring a feeling of wellbeing. It is suggested to play the diatonic scale starting from C on the piano (see chart of chakras and notes below). Listen intently to every note—feel it and sense your reaction to it. Does it feel good? Does it resonate within? Continue until you feel sure this is your note. Return each day for a week to find the note and sound it many times over. Do you return each time to the same note? If so, seek no more, THIS IS YOUR NOTE!

Remember, sound touches the physical body—it enlivens or subdues the mental and emotional bodies. When the three are balanced, the Spirit heaves a sigh of relief and dances within the sound waves.

I know a wonderful chiropractor who can help you find which note you are missing, and then, while listening to a tape of the musical scale (diatonic), you are restored to balance on reaching and hearing that note. I have personally experienced this, and yes—it was Note G! (see *THTCB* page 111).

Excessive mental derangement responds to the keys of F Sharp and A Major varied at intervals by F Major . . . the keynote of Sagittarius, ruler of the higher mind. Patients so suffering need the most soothing of string instruments, preferably a harp, and treatment rooms should be in restful tints of Green.[29]

—Corinne Heline

My present preference is that each note is associated with a particular chakra, or power centre, and has a colour connected with it; when these are brought together, they form a very potent healing tool.

Chakra	Note	Colour	Complementary Colour
1. Root	C	Red	Blue
2. Sacral	D	Orange	Indigo
3. Solar Plexus	E	Yellow	Violet
4. Heart	F	Green	Magenta
5. Throat	G	Blue	Red
6. Third Eye	A	Indigo	Orange
7. Crown	B	Violet	Yellow

What then is Sound Healing?

When an organ or body part is healthy, it creates a natural resonant frequency in harmony with the rest of the body. When the vibration of a part of the body is out of harmony, we have dis-ease and a different sound pattern is established in the affected part of the body. When sound is projected into the dis-eased area, correct harmonic patterns are restored, and there are many methods of healing with sound. Mantras and chants have been used for thousands of years. Many acoustic instruments are used in a variety of ways to effect change. The human voice is perhaps the most powerful musical instrument!

Crystal Bowls, too, containing the qualities of amplification, storage, transfer and transformation, are powerful tools in effecting change in one's life. It is important to approach the use of the Bowls with a reverence for the potential contained in combining intention with the properties of crystal.

It is now believed by many that "forced resonance" can be used to restore balance in the body through surmounting imbalanced conditions, if it is understood and used correctly. Every cell in our body is a sound resonator; every organ, bone and tissue has a healthy frequency at which it naturally vibrates. When dis-ease sets in, vibrations alter, creating sounds which interfere with the healthy frequency of an organ. We can change the vibratory rate of a body portion back to its natural frequency. Directing specific sounds through the chakras into the physical body creates balance. The entire energy system is strengthened.

Experimenting with vowel sounds and harmonics has been found to stimulate the brain and also the Pituitary gland—the body's chief and controlling hormonal (endocrine) gland just under the brain, often referred to as "leader of the endocrine orchestra."
—Olivea Dewhurst-Maddock

Vowel Sounds

VOWEL SOUNDS RESONATE WITH SPECIFIC AREAS OF THE BODY

Body Area	Sound	Note
Head cavity	I-I-I ("eye")	A
Bones in face/Stimulates 3rd Eye	A-O-M	A (above middle C)
Throat/Upper chest	E-E-E	G
Chest cavity/Body as a whole	AH	F
Abdomen (to navel)	O-O-O	E
Pelvis & Lower body	U (Oooh)	C' (above middle C)

What is Toning?

Toning is the process of allowing sound to move through you, playing you like an instrument (see *THTCB*, page 116). It is vocal sounding, in an ancient form used for attunement and healing. Toning is physically relaxing and helps activate your own healing energies; it focuses you into your centre and increases your awareness of who you are in body, mind and spirit. You can feel and hear yourself resonating.

Toning is a special time just for you. Focusing on your breath, posture and voice nourishes and restores you through the closest healing sound you have—your own voice! And it will help to improve your speaking voice, health and relationships. Be sure to relax as you tone—this is critical.

Ted Andrews writes:

> [Toning] is easily learned and applied by anyone who can speak. Simply, as we inhale we focus our mind on the region of the body associated with the vowel sound, and we sound it silently. Then as we exhale, we vibrate or tone the sound outward and audibly. . . . The breath takes the energy of prana and, combined with the vowel tones, they open specific regions of the body or consciousness.[30]

Sound will always resonate in the throat! We can heal by simply using the voice, by talking, using the vibration of Sound. Those who use their voices as channels for healing—also teachers, actors, singers—create a quality of In-tone-ation that touches strings of the heart, organs, emotions . . . and uplifts, balances, cleanses (by tears)—there are many ways.

Toning has two functions. . . . In addition to cleansing the aura, it invites the subconscious mind to co-operate with ideas held in the rational mind and to bring about a condition of wholeness, in body, mind and spirit. A whiny voice attracts lingering illnesses. . . . The Vowel sounds are most basic, because they open the centers of the body so that Divine Love or Kundalini energy may flow through.[31]

—Laurel Elizabeth Keyes

Case History — Chanting my name—Heather

As I meditated on the floor and the group chanted my name with love, my body felt as if it were being "ping-ed" by the tones I focused on. A few seconds later, I visualized my next phase in life (which I've been fighting for the last six to eight months).

I experienced peace, acceptance and understanding of this next phase. When I "came back" I was overwhelmed with the emotion, validation, loving acceptance and encouragement from the group. Thank you. —H.T.

The Tuning Fork

I wrote about Tuning Forks in my book *THTCB* (page 120). Olivea Dewhurst-Maddock also uses them:

> [The following] activities enable you to become familiar with the pure tones of tuning forks, their effects on body and mind, and their potential for healing.
>
> TOUCH TUNING. Activate the tuning fork, preferably middle C (256 Hz) . . . Experience its vibrations by placing the stem of the vibrating fork on the palms of your hands, soles of your feet, and crown of your head. Lie down and ask a partner to hold the vibrating fork's stem at various sites along the centre of your chest and down the back of your spine. You will probably feel certain sensitive "trigger points". . . Likely trigger points are the atlas bone at the top of your spine, and the upper tip of your breastbone, just below the hollow of your throat. At these sites, the vibrational energies of the fork are resonating and harmonizing the energies of your cells, bones, muscles and tissues.[32]

Try it for yourself. There is no wrong tone for the body, although some are more pleasant. The intent is healing and, of course, everyone is an individual with specific needs.

Remember above all that you are a tuning fork. Strive to register a pure clear tuning note. In daily life, be still and calm in your soul, so that you may vibrate in harmony with the music of the spheres. Thus you will be creating and not destroying the finer ethers in which you live—and live for others.

One of the great White Brotherhood, and a great musician, was the Master Pythagoras: to him the modern world of music owes much, for he taught his pupils about the soul harmonies, the heavenly planetary influences. He taught them wisdom and love.

—White Eagle

I have observed with interest, how people in cancer recovery can be part of a large group listening to the crystal bowls being played, the frequency will go directly to the affected person/s and the affected area where there is disease. The person will feel heat around the part of the body that needs healing. From my experience in working with many people in cancer recovery, I have found that those people have to change their diet to incorporate more live foods. We all should eat more living foods, sprouts, pulses, etc. created by the sun in order to get more photosynthesised energy. Also, it is important to remember our dream and live to actualize that dream. The unique quality of crystal sound as a form of vibrational medicine is that not only can the sound be played for the person in cancer recovery, but the whole family matrix can join. This creates a very positive support structure and emotional release for all.[33]

—Awahoshi Kavan

When people do not feel that their purpose is being listened to or actualised they begin to develop digestion problems or flatulence—because emotionally they sense that their purpose is being obstructed. . . . I would like to introduce the primordial sound of crystal into eating disorder clinics because I have seen it heal several young people that were diagnosed with anorexia and hospitalized for this condition. The complete octave of crystal sound quickly resonates the will to live and nourish the self again.[33]

—Awahoshi Kavan

Searching for your Soul Song?

Betina Lindsey writes in *Creativity Links* newsletter:

> I have read that in some cultures each person is given a song before birth. Parents sing this soul song to the unborn child and teach this song to all members of the tribe. Whenever someone is ill or unhappy or feeling bad, the entire village will sing the person's soul song to restore them to health. Many people testify to the power of healing sounds in their lives. My study of vibrational medicine has shown me that each and every person has a core sound, a vibration or set of vibrations that emanate from our being and body. Sounding these sounds restores us to balance, aligns our energies, and releases our creative life force. When we are diseased, our bodies vibrate in dissonance to our being. We can help ourselves return to health and balance by recalling and reinforcing our natural sound vibrations. You can sing yourself back to health. This is the essence of Sound Healing.[34]

A piece of music that touches the heart and therefore the emotions is Hubert Bath's Cornish Rhapsody. The music of Bach and Handel restores harmony—or use your hands to play an instrument to restore balance. This releases anger and frustration from the body. Great composers can: 1) accelerate chakras or 2) slow them down.

Let us try to understand healing: As we listen to sound, some healing will be entering the body; chakras can release effortlessly through the breath. Also, remember the holy sound "Aoum." When you sing it, aim to reach a point of sound coming through you, not of you.

Intention

It is the intention behind the action that determines the outcome. . . . (see *THTCB*, page 114).

All we need to do regarding intention is to state it out loud before Toning, sounding the Bowls or any combined treatment—for example: "My intention is to balance all my energies harmoniously and completely—and I do this with great love and understanding for myself."

The human body is an orchestra, each organ an instrument, complete within itself. Herein is to be found the basis for musical therapy. In the New Age, music will be the most important and successful of all therapeutic methods. . . . Harmony is the keynote of the spirit.[35]

—*Corinne Heline*

Be yourself. That is who you were meant to be. You are a note, a necessary note in a beautiful song![36]

—*Tolbert McCarroll*

Here's more from Dr. Willix:

. . . [A]nd choosing the right music for you, like choosing your mate, is highly personal . . . you need to be aware of how music affects you. . . . But research tells us that there may be universal reactions to certain music, e.g. in lab. studies, rock music can wilt plants. And it may wilt people too [see Dr. John Diamond's book *Your Body Doesn't Lie* (quoted in *THTCB* page 125)]. . . . Studies also show that music that is full of discordant notes . . . tends to produce confusion and irritability. . . . Heavy beats and strong bass vibrations also tend to overload the energy centres (or chakras) that govern sexual desire, and some researchers actually believe that heavy, loud, pounding music can lead to addictions or violence.

It has been found that the most relaxing music is that which has a beat close to a relaxed heartbeat—or about 60 beats per second. . . . Look for music that relaxes, releases anger, calms, cheers you up, helps you express emotions, gives you strength and courage, helps you focus mentally, or stimulates your creativity. . . . For example, patients with high blood pressure should listen to soothing music at least twice a day, once in the morning before they start their activities and once at night when they are starting to relax. . . . I recommend that patients with chronic fatigue listen to uplifting music that's in harmony with their guided-imagery work. . . . Every day we are bombarded with sounds over which we have little control, from jack-hammers in the street . . . [to] discordant vibrations of other people—"sound" that we can't hear but that affects us profoundly nevertheless. . . . Each of us needs to come up with our own sound prescription to counteract the negative energies of the sound around us. We need to pick up "good vibrations." . . .[37]

—Dr. Robert D. Willix, Jnr.

Steven Halpern found that his music stimulated the relaxation response better than Liszt's Liebestraum #3—a piece that music therapists and their patients report as "very relaxing." The results indicated that brainwave patterns, which changed very little during the Liszt piece, changed dramatically when the study participants were listening to Steven Halpern's Spectrum Suite. They relaxed immediately under the soothing influence of the alpha waves, and the right and left hemispheres of the brain balanced immediately.[38]

—Dr. Robert D. Willix, Jnr.

Try listening to Steven Halpern's Spectrum Suite (Halpern Sounds HS771, 1979).

Sound will be the medicine of the future.
—Edgar Cayce and Alice Bailey

In the United States, Dr. Andrew Weil of the University of Arizona has always foreseen the development of Integrated Medicine within the medical community. Such an evolution will be exciting.

Sounding your very Own Tone

There is a Sound in the Heavens, a special note, calling you . . . asking you to listen—listen—listen—listen. Your note is calling you to balance—balance your energies each day, to be in tune with all life.

How can we listen, and also tune in? The Sound of your note can come from within you—if you are still, and allow yourself to "Tone." Release whatever tone needs to come forth from deep within.

We need not be a singer, but simply a soul with deep faith, and in the silence begin to attune with our Higher Self—have intention (most important of all!) that we are Toning, tuning-in, to balance our energies. Then stand "tall." If there is pain, or wherever imbalance is felt, put a hand over it, make your intention in words out loud, then take a deep breath and as the breath is released, allow whatever sound needs to come forth, loud and clear for as long as your breath lasts. Repeat this twice more.

Together with intention can be added colour with the breath. Colour that part of the body—whatever colour you intuit. This adds to the healing quality. Tone several times a day, and note how you feel.

Each of us has our own tone or note, a specialness, a part of the whole "orchestra"! In fact, it is scientifically known now that each organ has a note, so we can say we have our very own orchestra! This is self-healing at its best, for at this point in the year 2001 and onwards, we are called to use Sound more for healing and wholeness.

Try it. Sound your note(s) and become one with the whole Universe. Make your melody and feel so good. Let your melody touch others as it "rings" true!

[We] have to learn to live through these energies which are being continually transmuted. Your body today is not the same as your body yesterday. Your mind and thought and philosophy have changed since a week ago. You are not the same person you were twelve months ago. You have grown, you have suffered, you have lingered with your regrets, your pride, you have had moments of intense pain and intense happiness; they are all energies moving through experience, transmuting afresh, transmuting the emotions, transmuting the love, transmuting the nature.

This is what life is for. It is constantly transmuting from one force energy to another in the perpetual motion of change and there lies happiness, there lies health, there lies security, in the acceptance of this continual octave within the experience.[39]

—Ronald P. Beesley

Your body is a self-healing instrument—if you give it a chance. It is genetically pre-programmed to heal itself. In my opinion, certain music heals by assisting the body to come into its natural state of balance and harmony. At this stage of research and development, it is both politically incorrect and legally irresponsible to state that a specific selection of music will heal a specific physical disease. . . . the common denominator in the vast majority of approaches acknowledges that the body heals itself most effectively in a state of deep relaxation. . . . Using music to evoke "the relaxation response" is one of the simplest and most effective ways of all—but you must choose the right music.[40]

—Steven Halpern, composer of healing music

WHAT IS YOUR NUMBER AND COLOUR?

We now need to get an idea about colours and planets and how they relate to notes and chakras and their purpose. Study these on the following pages and become aware of your place in all of this. It will help you to understand yourself more fully, to be able to "track" your movements as it were, and to make any necessary changes to enhance your progress towards self-understanding—and therefore to make your next "move."

Try this for yourself; we can then work toward a balance for everyone, and the world will be so much happier and calmer. It will be very peaceful . . . and we give deep thanks for that.

Always remember that you are watched over with great Love, and helped with insights along the Way. Remember too, to ask for help, for it is freely given if you will listen. So go your way rejoicing, Dear Ones. Life is for living, for forgiving, for loving above all—this includes your very beautiful self(ves).

So what is YOUR number and colour when you arrived on Earth?

OLD TRADITIONAL CHART

Colour	Number	Letters			Note
Red	1	A	J	S	Middle C
Orange	2	B	K	T	D
Yellow	3	C	L	U	E
Green	4	D	M	V	F
Turquoise	5	E	N	W	F#
Blue	6	F	O	X	G
Indigo	7	G	P	Y	A
Violet	8	H	Q	Z	B
Magenta	9	I	R		High C

Name:	R	E	N	E	E	B	R	O	D	I	E	Birthdate:
Number:	9	5	5	5	5	2	9	6	4	9	5	7/8/1920
Colour:	M	T	T	T	T	O	M	B	G	M	T	7+8+1+9+2+0=27
Note:	C	F#	F#	F#	F#	D	C	G	F	C	F#	2+7= **9=High C=Magenta**

Renee = 9+5+5+5+5=29, 2+9=11, 1+1=2
Brodie = 2+9+6+4+9+5=35, 3+5=8
} 2+8=10, 1+0= **1=Middle C=Red**

Jan Linch explains:

> The new numerological chart linked to the new geometrical chakras will bring about changes which your activated DNA coding will begin to attract. As you begin to experience changes in your life you will begin to attract new lessons which need to be worked through. You can analyse the colours which are linked to these changes by working out your date of birth and the full name which you feel comfortable using at this present moment in time.[41]

The following is the new Numerological chart. There are twenty-six letters (2x13) in the alphabet.

NEW NUMEROLOGICAL CHART

Colour	Number	Letters	Note
Red	1	A and N	C
Orange	2	B and O	C#
Yellow	3	C and P	D
Gold	4	D and Q	D#
Olive	5	E and R	E
Green	6	F and S	F
Jade	7	G and T	F#
Turquoise	8	H and U	G
Sapphire	9	I and V	G#
Royal Blue	10	J and W	A
Indigo	11	K and X	B♭
Violet	12	L and Y	B
Magenta	13	M and Z	C

Name:	R	E	N	E	E	B	R	O	D	I	E	Birthdate:
Number:	5	5	1	5	5	2	5	2	4	9	5	7/8/1920
Colour:	Ol	Ol	R	Ol	Ol	O	Ol	O	Go	S	Ol	7+8+1+9+2+0=27
Note:	E	E	C	E	E	C#	E	C#	D#	G#	E	2+7= **9=G#=Sapphire**

Renee = 5+5+1+5+5=21, 2+1=3
Brodie = 2+5+2+4+9+5=27, 2+7=9
} 3+9=12, 1+2= **3=D=Yellow**

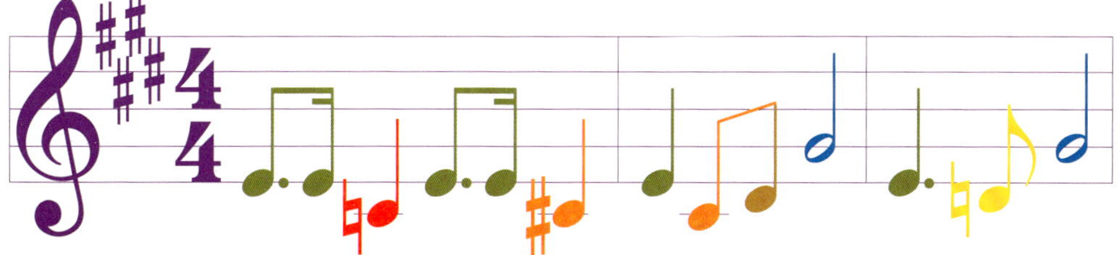

SHARRY EDWARDS—SOUND HEALING PIONEER

Another great pioneer of Sound Therapy is Sharry Edwards of Ohio, founder of Sound Health Inc. She has a "Bio-Acoustic Machine" that can generate very low-frequency sounds to train brain waves to produce missing frequencies. She can also create a voice analysis on computer to assess nutritional needs.

Jule Klotter, writing about Sharry in an article in the *Townsend Letter for Doctors & Patients*, states:

> . . . Sharry had unusually acute hearing; she was able to hear frequencies well above the normal human range. . . . [T]hese sounds seemed to emanate from the sides of people's heads, from their ears; and, each person emitted an unique frequency. . . . curiosity drove Sharry to learn more. . . . the Greek philosopher-mathematician Pythagoras . . . used music to heal. Like Sharry, Pythagoras reportedly heard sounds that others could not. . . .
>
> Sharry came to realize that frequencies emitted by the ears are *missing* from the voice. . . .
>
> According to data compiled by Sound Health, cancer patients are missing notes C and B. Nearly all multiple sclerosis patients lack notes D and A. Nerve disorders are accompanied by an absence of the note B. Eye diseases show up as a lack of note A, and depression appears as a missing or lowered G. . . . The note of G corresponds to the atomic weight of magnesium. . . .
>
> Sound therapy readily complements other therapies. . . .
>
> Impressed by sound therapy's ability to stop the shaking of a Parkinson's patient, officials at the NIH's Office of Alternative Medicine asked Sharry Edwards to write a chapter on sound therapy for its book on alternative therapies. Some progressive insurance companies are thinking about covering sound therapy treatments when used for pain management and rehabilitation.[42]

RESOURCES

1. *In the USA: Sound Health, Inc.*
   P. O. Box 416, Albany, OH 45710
   Tel (740) 592-5115; Fax (740) 698-6116
   Email <sound@frognet.net>
   www.soundhealthresources.com

2. *In Canada: Anita Van der Haeghe (trained by Sharry Edwards; has taken two levels plus Teacher Training).*
   P. O. Box 15555, 265 Port Union Road, Scarborough, Ontario MIC 2L0
   Tel (416) 284-6334

3. *In the U.K.: Elaine Thompson (trained by Sharry Edwards), Unicorn Centre, 57 Hillhead, Glastonbury, Somerset BA6 8AW*
   Tel: 01458 833238

There is also Bi-Com Bio-Acoustic Therapy originating in Germany—an advance on Dr. Voll's Vega machine.
Contact: Sylvia Lohs
2461 Bellevue Avenue,
West Vancouver, B.C. V7V 1E1 Canada
Tel (604) 925-1470; Fax (604) 925-5081

Music Played Publicly Everywhere

From where does the idea come that we must be assaulted by a certain kind of sound everywhere we go—in the supermarket, the restaurant, the hairdresser, the elevator, the bank, the department store—even in the washrooms? Who thinks that it is universally acceptable, that we become compulsive listeners to music not of our choice wherever we go? Who thinks that we just love it and are longing to have the accompaniment wherever we are, whatever we are doing?

Music is an art imbued with power to penetrate into the very depth of the Soul, imbuing man with the love of virtue.
—Plato

The simple fact is that sound touches our bodies, goes into the very cells. It can uplift or depress, or make one feel on edge. The secret of hearing music surely is to pay attention to what is happening in your body as you listen—does it produce energy, harmony, peace and balance?

Rock and Pop Music

John Diamond, M.D. writes of the harmful effects of the "anapestic" beat in some popular music, most notably rock and roll (though not all rock and roll. It is found, for example, in the music of the Rolling Stones but not in the music of the Beatles). This beat is against the heartbeat and confuses the natural entrainment process. When the anapestic beat is a part of the music, volunteers being muscle tested immediately lose muscle control and have to lower their arms.

> Using hundreds of subjects, I found that listening to rock music frequently causes all the muscles in the body to go weak. The normal pressure required to overpower a strong deltoid muscle in an adult male is about 40 to 45 pounds. When rock music is played, only 10 to 15 pounds is needed.[43]

Jean Houston describes her experiences with hard rock music as equally harmful:

> Right now, for example, with hard electronic rock, you get battered consciousness. Some kinds of rock music, at least as I experience it, have a toxic effect on consciousness. The reason I don't join an exercise club is because the music constantly played there is so terrible that I just can't take it. It may drive the adrenalin in the system, but it trashes everything else. This music is recapitulating the holocausts of the wars and the pogroms of the twentieth century, before the birth of an essentially new order of music, based on an expanded vision of what human beings can be.[44]

Meditation: Armchair Healing

Sitting very serenely in your high-backed armchair, visualize the great Sun overhead . . .
its beautiful golden rays pouring down upon you as you breathe gently in rhythm . . .
rising and falling . . . rising and falling. . . .
Feel the warmth of its rays on your body. . . .
With every inbreath, you know this healing warmth permeates every cell—
revitalizing, refreshing, energizing your whole Being. . . .

Take time to feel this radiance within . . . the utter simplicity of it . . .
the stillness and the Peace. . . .
After awhile, feel your whole Being longing
to help others to experience the same.
You CAN do this very thing from your armchair, simply and with humility.
Send forth into the Universe, from your whole Self,
thoughts of Love and Peace. . . .

See the rays of YOUR "Sun" beaming forth from your heart centre
to wherever they are needed at this particular moment. . . .
The loving, peaceful thoughts of one person
can change those of a thousand others. . . .

If every soul would take time to do this each day,
the whole world would be so Peaceful.

Chapter Nine

The Pure Tones of Crystal Bowls

At the present time the Hopi American Indian prophecy is being fulfilled with the *"Coming of the Rainbow People,"* through the keepers of the crystal bowls. This ancient wisdom has emerged to heal and uplift the consciousness of the universe through pure crystal tone.[45]

—Awahoshi Kavan

The Great Light within us is reflected in many ways, and one of the ways is through a crystal—so clear, so scintillating. The touch sets in motion many tiny cells, they dance, sending Light back to you, and within that Light or reflection is also colour and sound.

So it is with pure Crystal Bowls: the pure sound reverberates through the body, touching where it may and in so doing, creating balance. Next time you hear the sound from a Crystal Bowl, be very aware of the wonder of Nature, and how fortunate we are for these experiences which enliven us, and send us thankfully on our special Pathway rejoicing, filled with the Christ Light.

The Practitioner Crystal Bowl

Sky Holden Dunn, a sound healer practising in Salt Lake City, Utah, writes:

> The practitioner bowl is an all-quartz crystal singing bowl that has been designed with a quartz rod handle. This newly developed healing tool has been specifically created for therapeutic use for both the sound healer and as an addition to any alternative form of healing work.
>
> The practitioner bowl is a lightweight hand held Bowl that when played can be directed into and around any part of the body. This bowl is extremely resonant and will continue to sound long after a few seconds of playing. This lasting resonance gives the therapist ample time to direct the vibrations to where they are needed the most. . . .
>
> The practitioner [bowl] gives the user freedom to do multiple combinations of movement while the bowl is sounding. This is ideal for moving energy through the chakras and "sound massaging" the body. . . . [It] is available in many different frequencies and in all the musical notes. . . . very effective for stimulating the various

chakras throughout the body. . . . also ideal for cleansing the aura of the body. Sweeping the bowl around the body stimulates the outer layers of the auric field and produces . . . "a cleansing shower of vibrational sound energy."

The practitioner bowl can be used alone or in combination with other crystal bowls [and for "fine tuning" after all the other bowls have been sounded.] The practitioner bowl produces a very tight almost laser-like waveform that penetrates exactly where it is directed. . . .

The creation of the practitioner bowl is the creation of a true healing tool, specifically designed for sound healing.[46]

I have a Note D Practitioner Bowl for the Sacral Chakra; however, its sound touches wherever needed. It has been used to balance the energies in a sore knee—after ten minutes the pain dissipated. It has helped alleviate pain in a swollen hand; it can be used in many ways, as can each different note of the Practitioner Bowls.

In summary then, the Practitioner Bowl can:

- cleanse the aura;
- stimulate various chakras through sounding near them;
- clear the energy in a room;
- be used with toning; and
- be played at the same time as other Bowls.

Jack Brodie

Practitioner Bowl being played.

Jack Brodie

Frosted and Clear Practitioner Bowls.

Eva Rudy Jansen, in her book Singing Bowls (1990), documented the effect of Tibetan metal bowls; "It is possible to record the waves produced by singing bowls. It was found that among the wave patterns of different singing bowls there is a measurable wave pattern, which is equivalent to the alpha waves produced by the brain. These bowls, in particular, instil a sense of deep relaxation and 'inner space opening up'." I have seen this same experience described by many people following a complete session of the crystal octave.[47]

—Awahoshi Kavan

New Developments in Crystal Bowls

THE ULTRA-LITE FROSTED BOWL

There is now an ultra-light frosted Practitioner Bowl—a thinner quality frosted bowl—whose sounds ring very clearly and one can hear the sound of the octave; it touches the person very deeply. This Bowl looks like the Classic frosted bowl: it is lightweight, portable and can be played on the body or in the hand permitting more movement for the therapist with better access to chakra and meridian areas. Its tone is deeper than clear bowls, with single or overtones and continues to resonate for several minutes after sounding.

The Practitioner Bowl has the power to uproot deep-seated hidden emotional, spiritual energy blocks. . . . It pierces even the thickest walls that we as humans put up around us. It can be used with any type of meridian or chakra therapy and works very well with acupuncture, chiropractic adjustments, cranial-sacral work, massage or any other types of body-work.

—Dr. Gary Whitley,
chiropractic physician,
Salt Lake City, Utah

MINI PRACTITIONER BOWL

There is also a Mini 5″ Bowl to touch chakras above the head. It is a well-designed sound tool with an integral handle, and its range varies from 810 to 860 cycles. It is suggested that this bowl be used in conjunction with lower-pitched bowls.

Fabien Maman, founder of Tama-Dō Academy of Sound, Colour and Movement, and author of four books including The Role of Music in the Twenty-First Century, gives dramatic accounts of the healing effects of sound on cancer cells. He uses various acoustic instruments and notes as well as the human voice, and incorporates colour and movement in his work, which deals with the subtle bodies, where dis-ease is created.[48]

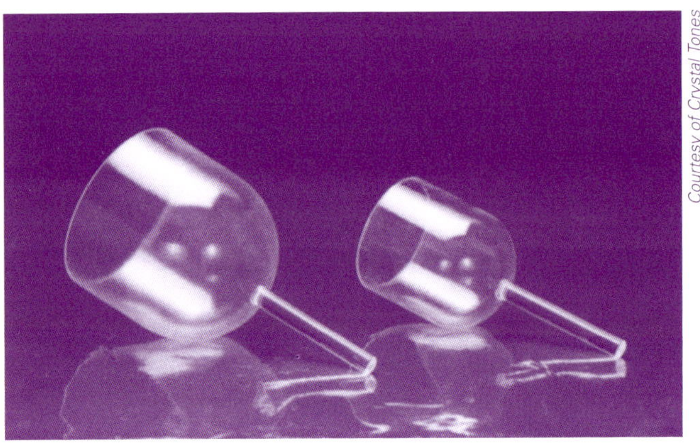

Courtesy of Crystal Tones

Practitioner Bowl and Mini Practitioner Bowl

THE TUNING FORK PRACTITIONER BOWL

This is an ultra-lite frosted bowl 4" in diameter by 7˘" long with a range of 850 to 900 cycles. It sounds a high frequency and can be used in conjunction with the Mini practitioner bowl to produce profound effects.

I have worked with many cancer patients; they are sometimes very sick and in the dying process . . . they report feeling less anxious and more peaceful after their treatments with sound (of the Bowls) and Light . . . also more energised. They all seem to experience peace and calmness . . . and are more accepting of their journey with illness and possible death.

—Barbara Jensen, sound and esogenic colourpuncture therapist, Park City, Utah

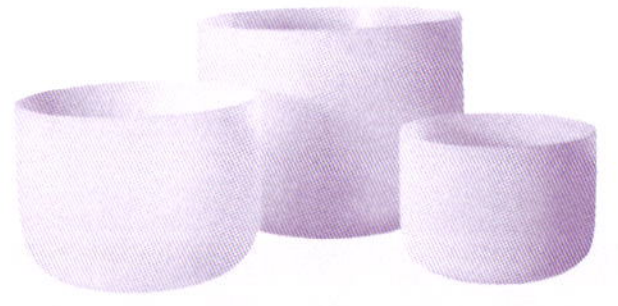

For more information on where to obtain crystal bowls, consult the author's website:

www.angelfire.com/ca/crystalbowl

Meditation: On the Wings of the Wind

Become for the moment a gossamer-like head of a dandelion gone to seed,
one that travels on the wings of the wind . . .
is so beautifully and intricately patterned . . .
so soft, downy and silky. . . .

Feel yourself to be in the heart of the wispy pattern . . .
and let the wind take you on its wings . . .
higher and higher, beyond the treetops into a summer sky,
where all is tranquil and warm. . . .
Feel safe within its intricate beautiful pattern—
and from this point, let your gaze go far. . . .

Ahead is a great silvery-blue archway filled with rainbow light. . . .
You are drifting towards it . . . almost as if drawn there by an Unseen Power . . .
and as you float through the archway,
you join myriads of other gossamer-like Beings
whose Love enfolds you—
till all become part of the One beautiful pattern woven by a Master Hand . . .
and you are at Peace. . . .

Chapter Ten

Colours of Life

> There is an affinity that comes through our feelings, for even as we link to every shade, design and colour upon Earth, so our heart joins to the creative spirit within. . . .
>
> —Ronald P. Beesley

So What is Colour?

The Sun gives forth emissions of particles and wavelengths. We in turn have been given the gift of sight, to interpret and convert these particles and wavelengths into colour, thus turning the earth's grey forms into a profusion of colour that permeates and transforms our Being, bringing life and joy into our lives.

When light enters the eye it strikes the special cones at the back of the retina, where the wavelengths are converted to colour and sent on to the brain. These in turn trigger the release of hormones that can influence moods and activities such as heart rate and breathing. We also receive colour through the etheric aura (home of the chakras). The importance of colour cannot be overstated—it can have a deep psychological effect on us.

By our ability to take the life-giving energy from the Sun and convert it to colour and healing energy, we can uplift our spirits and transform our lives. Refraction by a prism reveals all the individual colour components of white light that can be used singly or in combination to heal our bodies.

- The Sun's energy is essential to our wellbeing, bringing us colour and light to sustain us; without it we cannot exist.
- Colour is omnipresent. We enjoy in Nature spectacular sunrises, sunsets and rainbows, and absorb the greater energy available at those times.
- Colour influences our attitudes, feelings and thoughts, for example love = Pink; anger/excitement/passion = Red; studiousness = Yellow; peacefulness = Blue. Colour has enormous influence on our lives, often manifesting itself subconsciously at a deep soul level.

Those who have studied colour in depth say that colour (as well as sound and fragrance) will be very important in the Age of Aquarius, that we will discover how vital these factors are, and that colour feeds us on every level of our Being.

To bear fruit on all the levels—physically, mentally, emotionally and spiritually—we have to use all colours. How you feel about colours tells much about yourself. Do you have an aversion to Orange, for example? The best state is where you are comfortable with all colours and have no aversion to any colour. You can then use the full spectrum to advantage.

Master Djwal Khul has said that in the current decade, the knowledge of colour will expand greatly, and that you and I are the pioneers. A deeper understanding will evolve. "The Aquarian Age stands for colour, sound, fragrance and movement."

Sometimes colours can make us feel uncomfortable. Clothing materials act as filters, so our body receives colour. Drab, dingy colours harm our spirit, morals and health, and give us an inferiority complex. It is very important for mothers-to-be to wear lighter colours.

We are here to spiritualize matter, and Light cannot come through us until we are clearer prisms. Dis-ease manifests when we are cloudy. Our purpose is to become the perfect rainbow and we have a daily choice, every morning.

Jeanne Elizabeth Blum, in her book *Woman Heal Thyself*, reports on the use of colour for healing:

> . . . Dr. Norman Shealy of Springfield, Missouri—a noted neurosurgeon—has a healing facility in Springfield. One of his treatment methods involves using sequenced flashing colored lights, which balance the neurochemistry in the brain. He works with patients who have chronic pain, depression, sleep disorders, and other assorted problems. As a Diagnostic Intuitive (meaning intuition figures largely in my healing work), I studied the *science* of intuition with him at this facility and found his methods to be fascinating. In many hospitals across the country, leading surgeons are working with intuitive diagnosticians who mentally scan the body for illness and disease. It is now possible to acquire a Ph.D. in this field of study.[49]

The purest and most noble minds are those which love colour the most.
—British critic John Ruskin

At the height of the Egyptian civilization, priests linked into the highest aspects of the Red and Yellow rays—representing love and wisdom. Today we are working through the Green rays, to bring harmony and peace. Soon humanity will move into the knowledge of the deeper rays of Blue—for the soul. There will be peace as we work our way up the ladder of colour.

St. Martin's Lane Hotel in Covent Garden, London, in the Theatre District (owned by Ian Schrager), offers rooms with a full range of lighting—each person can choose the colour in which they wish to be bathed!
—USA Today, October 2000

All Life is Colour

Wherever we look we see colour in its many shades. And just to take a walk in Nature is to walk in natural hues . . . to be calmed by the blueness of the sky and the sea . . . to be made to feel balanced by all the different shades of Green: in the grass, in the leaves of trees and flowers and bushes. If we could study the colours and shades, even basically, we could have in our mind's eye all the natural therapies we need, for Nature is so healing.

We need Red sometimes for extra energy—a pure rich red like that of poppies, tulips or the deeper red of velvety roses and even leaves with reddish hues. These colours we can inbreathe to diffuse through the cells of our body.

What of Orange? To give us good appetite, for one thing; to open our minds to wise thoughts, to give us the urge to seek for Truth and Wisdom and also to give us a warmth. And then, as more light is added to the Orange, we have sunshine Yellow, a wonderful tonic, an upliftment in our hearts, almost like the warmth of the Sun on our body, but as well, a great cleanser, cleansing bodily toxins in every possible way. Yellow foods are very cleansing. We see that everything in Nature is there for us to use if only we realize how much we can partake of the God-given therapies.

Colour is so important and has extraordinary meaning in our daily lives. Tinge the Yellow with Blue and we have all shades of Green again, to calm us with the qualities of the forests and glens, the Sun shining through the leaves. It is wonderful to have our home full of vibrant plants to help us open our hearts to these beautiful vibrations of tranquillity. The Green emanates this calm in response to our loving care.

In all creativity we need Blueness, and with a touch of added Red and Brown to make Indigo the Blue becomes a rich, rich colour to help us achieve creative talents, a kind of rainbow deep within. Indigo has many enriching qualities—it helps us expand our inner vision to see beneath all things with a greater understanding, to find only the needs and the true ideals in the simplicity of life.

And out of the deepness of this wonderful colour of Indigo comes the palest Amethyst and all shades of Lilac—a very healing vibration. Its softness helps us to integrate all we know and realize; it seems to illumine our thoughts and make them clearer. It is like soft chiffon waving, enfolding us, breathing tenderness within and around us, a greater understanding that all life is a Oneness and we are part, not only of the Great One, but of each and every one. There is greater compassion as we are enveloped in this pure colour of the amethyst crystal—able to add to each others' vibrations and make ourselves healthier . . . but also Mother Earth a beautiful place to be.

Let us use these colours in every possible way in our daily life, and learn as much as we can—use the knowledge intuitively, with the clothes we wear, the colours in our homes and our work surroundings and in our food and our thoughts. Walk in a Pink "cloud nine," thus happily radiating unconditional Love to all with whom you come into contact—and light their lamps for them.

Colour Therapy is Soul Therapy

Colour is very important. There is immense power in colour—therefore a capacity for healing. Colour has a deep psychological effect on us; in exposing the subconscious mind to *pure* vibrations of colour (and sound), one balances the neuroendocrine system.

All nerve impulses from the legs, arms and torso pass through the spinal cord to the brain and head, so the throat and back of neck are very important! Let us remember that the chakras are ports of exit and entry for thoughts/energies, and the healing of the body is really purification of the blood.

The practical healer is actively involved in consciously reconditioning himself/herself, cooperating with and understanding the laws of Nature. And let us remember—the higher vibrational energy always feeds the lower vibrational energy, so one loses in a crowd of less evolved people, and in marriages; and that the act of forgiving oneself polarizes one's own electrical field.

Sound is within colour, and we hear magnificent tones as we view colours. If we do not hear them they are still there, deep within, and the colours trigger the sounds.

—White Eagle

Soul Colours—Another View

Suzy Chiazzari suggests:

Choose three colors:

1. The color you find the most *pleasing*.
2. The color you find the most *relaxing*.
3. The color you find the most *inspiring*.

The most pleasing color is your "soul" color and this color preference usually remains the same over a long period of time. You are drawn to your soul color intuitively and this color will represent your personality and outlook on life.

The second color you chose is your color of relaxation. It is the color that has a calming and soothing effect on you, bringing you healing on a physical, emotional and spiritual level.

The third color is inspirational. It is your mental color and helps you aspire to your higher self.[50]

It is more important now than ever before to use *both* colour and sound with complementaries in order to achieve an harmonious balance. Perhaps the undercharged chakra needs the key colour, and the overcharged centre needs its complement.

Working with Complementary Sounds and Colours

In order to achieve a balance of energies, we need to put together the complements of sound and colour (see *THTCB*, page 103). With sound, musicians know that the fourth note above any given note is the complement of the original note (the interval is called a "fifth"). For instance, four notes up the scale from Middle C gives G. To tone these two notes every day would be very balancing in a healing energy way.

In terms of colour, this would give us Red for Middle C and Blue for G. To alternately tone these two notes while visualizing the colours Red and Blue would give a feeling of goodness, calm and wellbeing.

Remember always that balance is the aim in healing; we are balancing the energies of the etheric body.

When we work with the Etheric Body, always remember that this energy also affects the other bodies—especially the Astral. These days we are being flooded with special light by the Light Beings, to raise our vibrations. This affects our Astral body greatly, causing much imbalance; it is not easy to live through. But if we keep our aim high, and remember that we *are* channels for that beautiful energy, this helps us to move peacefully through each day.

Try different combinations for yourself. Choose a colour to which you are greatly attracted today, and work with the complementary colour and sound. Test for yourself how you feel, how you react. Record it each day; you will be fascinated with the results.

Once you have worked with this yourself, you can help others to understand the power of colour with sound.

As you tone, try using different words or mantras, such as AOUM or I AM THE LIGHT, or simply I AM. Make your very own mantra. Notice what effect it has on your body, then try the short meditation at the end of this chapter to begin your new day.

COMPLEMENTARY SOUNDS AND COLOURS

Colour:	Red	Orange	Yellow	Green	Blue	Indigo	Violet
Note:	Mid C	D	E	F	G	A	B
5th Interval:	G	A	B	C'	Mid C	D	E
Comp. Colour:	Blue	Indigo	Violet	Magenta	Red	Orange	Yellow

History of Spectro-Chrome

The Egyptians used colour for healing in their temples. Before them there were many others. The Rosicrucians have successfully used colour for healing since the 15th century.

Dr. Dinshah Ghadiali (1873–1966, born Nov. 28 [Sagittarius], in India), was a Zoroastrian (Religion of Light). Dr. Ghadiali's favourite quote was from the Bible: ". . . and God said, 'Let there be Light.'" He was inspired by the work of Sir Isaac Newton, German optician Joseph von Fraunhofer, Seth Pancoast (who wrote about Blue and Red light) and Dr. Edwin Babbitt. Dr. Ghadiali, an electrical engineer and a fluent linguist, also studied medicine and was a Fellow of the Theosophical Society.

In 1896 he visited the USA, where he met the inventors Thomas Edison and Nikola Tesla. In 1897 he used light (through an Indigo glass bottle) on a friend's daughter with mucous colitis—she recovered after three days.

In 1911, Dr. Ghadiali emigrated to the USA, at which time he dropped the family name of Ghadiali in favour of simply "Dinshah," and the Dinshah Health Society was formed. It is now headed by his son Darius. A few years later he created Spectro-Chrome. This is a method by which colour can be radiated to specific parts of the body by means of coloured filters placed over lamps (not less than 60 watts) for a certain length of time, with many beneficial results.

In 1920, Dinshah began lecturing on Spectro-Chrome—there were many lawsuits and he was imprisoned several times for using colour healing. We must realize we all benefit from his 67 years of research and his sufferings, for now we can use colour for healing. We are learning from his experiences, and are supremely grateful. In 1966, on April 30, Dinshah left the Earth.

Dinshah's work was recorded by his son Darius in the book *Let There be Light*. We recommend the best way to use Spectro-Chrome is to read his book thoroughly; it gives explicit direction and many explanations. There is also a Colour Therapy Home Study Course available from the Colour Institute of Canada, by Audrey Ann Lowrie. (I have completed this course; it is comprehensive and highly recommended).

CONTACT:
Dinshah Health Society
P.O. Box 707
Malaga, NJ USA 08328
Tel (856) 692-4686
www.wj.net/dinshah

CONTACT:
Colour Institute of Canada
P.O. Box #59
Perkinsfield, Ontario
Canada L0L 2J0
Tel (705) 549-1999

Treatment by Light

In Europe, colour-therapy is considered to be new, but colour treatment, and treatment by light, is as old as the hills. It was used in Atlantis . . . also practised in Egypt. Even today, colour-healing is used in remote parts by tribes which are remnants of great civilisations and which retain certain knowledge handed down by their wise men. The ancient Sun people drew upon cosmic rays by the mental and spiritual power of the healer, and by pouring selected colours into the patient's body under the direction of the healer's will. You yourselves are beginning to awaken to the importance of colour in your surroundings, in your clothing, and there are advanced healers who understand how to apply certain colours, by use of material agencies, to selected parts of the body, with notable results. Later this same application of colour will be used by the *spirit* of the healer, who will understand how first to absorb and then direct the colours that the patient needs.

Yes!.....I am catching the thought! Coloured lights, coloured lamps, are effective, and have their place. But when you can get behind the physical or material representations of colour, and can draw upon the cosmic rays and the sunlight, as did the ancient brethren, then minor disease or "inharmony" in the physical, etheric or mental bodies will be healed.

Let us then try to understand how great a gift is light! Sunlight, the white light, can be subdivided into the seven primary colours. The seven colours are permeating every form of life on earth, and each has its vibration; even the herbs used for healing, according to the influence of the planet which causes their growth, reflect certain vibrations of colour. It is the vibrations from these herbs, when introduced into the physical body, which cause the readjustment from inharmony to harmony. Disease is lack of ease, lack of harmony.[51]

—White Eagle

A Clear Blue Sky

Overhead daily there is a clear Blue sky, even if it is not always seen, covered by grey clouds or rain. But like the Sun, it is always there, and so we know for sure that we can count on this beautiful Blue "umbrella" to spread its heavenly colour above us, all around us, and give us a feeling of absolute Peace if we will only look for it—Peace all around us . . . and above all, to infiltrate so that we feel it within, deep within our hearts.

We encourage each of you to be much more aware of this Blueness, of this wonderful comfort from the very heart of Nature, to feel encompassed by its arms, to know that within the Blue Heavens there is this great assurance and wonder of the great Peace we can actually encourage inside ourselves.

See the Blue then as a wonder of all Creation, as a beauty to behold, and as a comforter (which we all need some or all of the time). And it is free, there to behold, gaze upon, drink in deeply as we inbreathe its glorious peaceful colour and feel blessed, so very greatly blessed, and give thanks for our daily Being.

Meditation: The Blue Heavens

Within the Blue of the sky are so many lovely qualities of healing . . . of nurturing . . .
of stillness and Peace. See yourself now lying on the grass . . .
cool green grass . . . your body touching the earth . . .
and above you the great Blue expanse of the sky . . .
not a single cloud . . . such an expanse . . .
wherever you look is Blue!

And with the warmth of the Sun on your body—
the earth energies beneath you nourishing and wrapping around you—
allow yourself to be in the sea of Blueness . . .
beyond the horizon. . . .

All cares fade into the Blue . . . it melts into you . . . until you feel part of it . . .
the Blue of the Heavens—never-ending, all-encompassing, healing, soothing—
filling your heart centre with such gentleness, warmth and compassion
for all fellow creatures. . . .

Nothing else matters—BE STILL in the Blue Heavens. . . .

Chapter Eleven

Red—The Blazer of the Trail

> The lesson of our age is that the capacity to heal ourselves, one another and our planet, lies within us. . . . Colours heal by the spirit.
>
> —Author unknown

The Blazer of the Trail, the courageous one, the leader of the pack, the fiery one, the one of anger, of "do and dare" are manifesting the Red energy. As Red fills one, it raises expectations, raises one's hopes towards a completion.

These marvellously stimulating energies can be yours. Simply envision the Red energy, breathe it in from deep within Mother Earth. Allow it to climb the ladder of your limbs toward the pelvic chamber, to fill you—gird your loins, as it were, set you afire with courage, deliberation, concentration and a great upliftment to accomplish whatever is needed for the moment.

Infants love Red. It grounds and energizes them, coming as they are from a plane where everything is Light. If they become too energized, use Magenta. Pink is also a possibility as a relaxant (Pink is part Red).

Red warms the body. People with cold hands and feet should wear Red gloves and socks. Wearing Red at parties replaces the need for alcohol. Too much Red is overpowering—people will try to manipulate.

Red is excellent for anaemia, arthritis (for acute or inflamed arthritis use Blue) and the arteries. It's good after childbirth, enlivens the depressed, neutralizes fear and helps to ground a person.

Red has the heaviest and slowest Hertz rate (rate of vibration) in the spectrum. Many ill people need this basic, hot energy. They will experience heat going into the body as a quality of power in the spine. DO NOT USE IN CASE OF HEART TROUBLE.

The resonance of the Red note C played on the Crystal Bowl can fill you, move all the cells in your body so that they can come into a balanced state and you feel great within. Remember to add the complementary note G to complete the balance. You are encouraged then to be aware of all these possibilities.

On Purifying the Bodies

Here is something on which we can be mindful, something we can do daily for ourselves to work towards a shining edifice for all to see. How can we work each day to purify our bodies? What special "knack" is there to this? Can we do it just spiritually? Or does it have to be a combination of mind over matter plus spirit? Where do we begin? Does it take all day long?

Try to picture for yourself a set of Russian nesting dolls, where each "egg" lies within another, starting from a very small one, getting larger layer by layer. Then think of the tiniest egg. See it merely as a seed within the heart, waiting to be nurtured so that it can grow larger, and eventually flower in all its glory. This needs nurturing with meditation, quiet moments, loving yourself and listening to your heart too, hearing what you need to do just for the day.

Then, having nurtured the seed, see it enfolded by a gorgeous pale Pink egg, signifying the emotional body. How Pink is it? Does it need more Love? Is it overflowing with Joy—paler Pink—or is it rather Red, with upsetting feelings, or is it Grey with depression?

Work on it, give it understanding and more Love, until you notice it is Pink and beautiful and shining.

Next there is a larger Yellow egg enfolding those two. It represents the mind—all your thoughts and how you arrange them. Are they neatly packed within the Yellow egg? Or are they spilling all over the place—no order, no pattern, no balance? Can you get a "handle" on them? Discard what is unnecessary—keep only the good thoughts. Marshal them in wonderful order so that they can only produce more lovely thoughts. Leave the drab ones, the Grey and Black ones; allow them to drop away and withdraw into the mists of time. Let your mind be like the Spring—full of lovely new inspiring thoughts that can be transported by the ethers, touching others in that journey and giving inspiration to each.

And then the last, larger egg—the physical body. It can be a wonderful palish, misty Blue quality—shining, effervescent, spreading its wings, encompassing many. Work hard daily to purify your body—with deep breathing, with good, pure thoughts and thankfulness as you prepare food and then eat it. Allow it to nourish, release everything not needed, not only through normal channels, but also through the skin. Be aware. Love your beautiful body—it is a great work of art and you are continually recreating it. Do your work lovingly. Use every creative ounce within you to make and keep it pure, sparkling and habitable, so that you are a shining Being, radiating pure Light through the pores of your skin, and above all, through your eyes, the Windows of your Soul.

Chakra #1—Root

The first chakra is associated with survival, vitality. It is felt as fire that will warm the legs, spine, senses, the whole of the body.

Positive Emotions: Calmness, stillness, quietness, removal of tension.

Chakra #1 works together with #8 (Gonads and Crown), to produce hormones used by the adrenals to help control allergies and infertility, and to keep us aware and alert. Our energy reserves are here.

Malfunctions: Carries anger, fear, fixations, stress.

Imbalances: Anxiety and hyperactivity, obesity, hemorrhoids, sciatica, constipation, anorexia nervosa, degenerative arthritis, knee troubles, adrenal exhaustion.

Many have an impacted lower spine (coccyx and sacrum). The muscles of the rectum are tense, and this produces congestion and hemorrhoids. Lack of muscle tone in the lower bowel and abdomen occurs because ligaments are either too tight or too slack.

AS THIS CHAKRA OPENS

When Chakra #1 opens, there will be a general feeling of wellbeing, a warmth, better circulation, a better understanding of the sexual impulse and the actual act, a regenerative energy. Being stuck in this centre means a concentration on basics—sex, passion, survival. As the centre begins to open wider and balance, then moderation is experienced, together with a healthiness. The Kundalini rises until all chakras are energized and vibrating with glorious light, sound and colour. This comes about chiefly through an active life and devotion to some form of practical service.

The energy of the Earth is linked with the Root chakra. Our Soul's lesson is to learn to be in service. Earth energy gives souls the patience, courage and tenacity needed in the search for truth, to learn to obey the laws of health and harmonious living.

We all need some Red:

- It is the greatest of all colours for stimulation;
- It gives shy people confidence;
- But anger is also a Red energy that can disturb the energy field of everyone in the same room with an angry person. It leaves energy in the space which can upset everyone's auric field for the rest of the day.

Those with a deficiency of Red lack confidence:

- have a fear of being abandoned;
- feel weak and have little interest in sex;
- worry a lot.

It is suggested to drink cranberry juice. Also, walk barefoot on the grass early in the morning or walk on a Red carpet.

Grounding and Cleansing

Use Red to ground; visualize it in the pelvic area, send it down through the feet into the Earth. Play Note C (Red). We need to tone Middle C and play that Crystal Bowl for the Red Chakra. This is excellent if we alternate with its complementary note—G—visualizing Red at the same time. Also:

- Walk with awareness, feeling the flow of energy to the feet;
- Discharge negative energies through breath; have a sense of the Earth attracting the discharge;
- Wear a crystal—Amethyst is good for absorbing negative energy;
- Appreciate nature—work in a garden, cut wood, see animals. Remind yourself to do these things until they become habit and they come naturally.

The kiss of the sun for pardon,
The song of the birds for mirth.
One is nearer God's heart in a
Garden,
Than anywhere else on earth!
—Dorothy Gurney
(1858–1932)

Case Histories with Crystal Bowls

Bowls G, A, C, D, E, F, F# and B were played—I felt them in ears, neck, head, heart, Solar Plexus and Root. I felt a great melding of vibration—stress and tension went away, and I felt peace. —P.T.

M. needed help as his lymph channels were clogged and he was on a fast to help clear them. Red light was used on his feet and while lying there quietly he saw his body from outside himself. He cried—it was a wonderful experience, he said. Afterwards he felt very balanced, and his feet were warm and toasty. —M.W.

Debbie Cottle, a massage therapist and channel for healing in Salt Lake City, Utah, says: "All people resonate with the C Bowl more easily, and starting with this lower harmonic has a more positive influence when bringing in a higher tone. Always begin with a C Bowl to start things resonating a little bit slower and then move upward as the person can tolerate the chaos being created to allow clearing to occur."

She continues: "I have used the C and G Bowls more than any others. I feel they are the most powerful combination, balancing the throat, which is the most physical of the spiritual chakras . . . allowing for the manifestation of spirit into this dimension through the voice and the root chakra. It is vital to have alignment of these two chakras to manifest spirit into the physical, and when these two are combined they create the sacred purple hue."

Pink

Pink is White Light introduced into Red. It lifts, is more caring, softer—a gentler and faster vibration. Pink is linked to God's divine Love. Surrounding people with Pink breaks up crystallized thought forms. See yourself and others in Pink—it transmutes negative conditions around a given situation.

When we worry a lot, it's good to introduce lighter Reds. If someone is very irritable, use Rose Pink—it helps the nerves, feels warm within, relaxes. Irritable people are starving on many levels. Use for family quarrels, a sick world, a sick animal, relationships. Soft Salmon Pink represents universal Love for Humanity.

In the book *Colour Breathing* by Linda Clark, a woman colour-breathes in Pink light for a few months. Her wrinkles disappear and she looks years younger; people don't recognize her. Pink light bulbs can be soothing in a room. Pink light can be sent over the phone, as absent healing.

Pink is the winning "energy of Love." It is very effective for evening meditation, for forgiving others and ourselves. Use Pink energy for your personal love note. The more Pink you give, the more you have; it can never be used up.

Those with a Pink aura are in love; pregnant women often radiate Pink: their attention is away from themselves onto another. In relationship, you are loving the Divine Mother within the woman; a woman loves the Divine Father within the man. Pink attracts beauty and goodness. It is now used in geriatric, adolescent-crisis, family-therapy, prison-reform and business environments.

Music for the Root Chakra

Suggested music for the Root Chakra:

- Brahms' Symphony No. 1 in C Minor.
- Ave Maria on *Gregorian Chants Vol. II*, sung by the Benedictine Sisters of Mt. Saint Scholastica with sounds of nature (e.g., waterfall) as background. Not only does it have the impact that Schubert's original had, it also has the power of Gregorian chant behind it, resonating the colour Red for the Base chakra—responsible for our groundedness and our basic needs—in the key of C. This piece of music is said to correct for egocentricity, perhaps through learning to become connected—part of the whole. This is the opposite of being absorbed in ourselves, to the detriment of others—humans as well as animals, the plants and Earth.

Whither Thou Goest—There Go I Too!

It is a fact that the Angels, Shining Ones, Guides, God-like Energies (whatever you like to name them) DO walk with us; many of us will not be consciously aware of this wonderful partnership, but there are some who walk the Earth with the sure knowledge that we are accompanied by the invisible (to us) Beings who enfold us with their never-ending Love.

Such a comforting thought! Picture then, for yourself, a wonderful pathway stretching ahead, very straight, country on either side, fields, nature, some trees, and above them the golden Sun pouring down. And you are on this pathway, sometimes gazing directly ahead, but more often looking from side to side. Sometimes you are distracted by a beautiful tree on one side, taking time to sit at its feet, other times turning to the opposite way, finding a ditch with muddy waters and wondering how we can keep to the straight and narrow without falling into the ditch.

All the time there are highlights and shadows when the Sun disappears for awhile, but the Shining Ones are there too, matching their footsteps with ours, silently, lovingly with deepest compassion and loving helpful thoughts hoping that we are listening. They walk with us constantly, hoping to hear our questions on life, our asking for help. Our longings deep within our heart centres are known to them, so that these can be brought to the surface, looked at and acted upon.

These are the stirrings which are looked for, hoped for and encouraged so we can then begin to walk in the Light, with more understanding, and above all, releasing the abundant JOY we have until now perhaps locked deeply in the heart centre. Let out the Joy then; release all the doubts and fears, instead letting the Love flow greatly to yourself and THEN to others around you. Allow this Love, which is Light, to join with the great Love which enfolds you from your Higher Self and your Angel guides. Walk in the Light now, peacefully, and know that the Master's radiance enfolds you.

Suggested Music for Anger

- Beethoven—Egmont Overture
- Tchaikovsky—5th Symphony—last movement
- Brahms—Piano Concerto #1
- Scriabin—any work
- Handel—Harp Concerto in B♭
- Schubert—Prelude to Rosemunde

Meditation: Journey in Colour

As your breathing becomes calm and steady,
enter the Stillness and mentally travel to the foot of a tranquil mountain. . . .
Here amidst the green of nature . . . birdsong . . .
and the scent of wooded paths is all quietness. . . .
You sit for a moment's rest.

Leaving behind your physical and etheric bodies, also your mental and emotional sheaths
(represented by your Red-tipped Yellow slippers left neatly at the foot of a tree) . . .
in your Body of Light, begin to rise towards the mountaintop and beyond . . .
first in a pale Green, wispy Light (the Astral body);
see it floating upwards like soft chiffon.

Halfway up the mountainside, the pale Green changes to a beautiful Blue (your Soul Body)
and continues floating upwards . . . till above the snowcapped peaks
it changes to palest Lilac (your body of pure Spirit).
Go with it . . . beautiful, iridescent, shining Light . . .
to wherever it takes you . . . into glorious Light and Peace.

When you are ready to return to "Earth" . . . float back to the mountaintop
where your light changes to Blue. You find yourself coming gently and easily down . . .
down, until you notice that your Light is now Green—a beautiful strong colour
. . . and further, until you suddenly find yourself in your physical body,

slipping into your Yellow slippers with the Red-tipped toes,
fully back to Earth but recharged with energy
and filled with a wonderful Peace.

Chapter Twelve

Orange—The Spirit of Health

Make everything in your life creative, whether it be poetry, music, painting, sculpture, writing or inventing. Unify all forms of art to create a total sense of beauty, form, uniqueness and elegance.

Concoct radical ideas and theories—the more radical the better, then try to disprove or corroborate them. Go off in completely new directions, discard reason, be bold, ridiculous and absurd to break from the usual stream of consciousness.

Remember, there are no limitations to creativity unless you yourself impose them! The field is infinite, the scope limitless. There are no laws to be broken, only man-made conventions to which we are not bound. Put aside your past learning, wipe the slate clean and start again using inspiration and insight rather than memory.

When creativeness is applied to music, art, poetry and writings use beauty, flare, style and total abandonment as your creative medium.

—Jack Brodie

As We Think, So We Are

As you arose this morning, what did you think? Were you aware of the glories of the morning Sun? Did you feel a glow inwardly that something wonderful would happen today? Do you think that you are a very lovely soul? Did you feel you could accomplish much? And be so grateful for God-given energies? That you could spread them around to your family and fellow beings?

Or were you in a depression, dreading the new day and what might unfold for you? Were you already feeling miserable and so unloved, that there is no one around to comfort and uplift you and your thoughts? That you were deeply into fear and unlovely things?

Here is a great opportunity for each and every one, to decide

Listening to music should bring out everything that is best in us. It should be like wind in our sails bringing our ship nearer to our heavenly predestination.
—Omraam Mikhaël Aïvanhov

HOW we are going to be on awakening, what the day is going to offer us, how we can overcome thoughts of dread by concentrating only on the good, the true and the beautiful. How can we do this for ourselves? How can we set the "tune," the mood, by ourselves? How can we endeavour to be in "top form," without thoughts that go below the line of fear and unkindness, lacking Love? What shall we say to ourselves?

Firstly, as we lie in bed slowly waking up to the day, let us give thanks to the Great Energies of the Universe, that we are here and ready for the brand-new day. Let us picture in our mind's eye the Sun in all its shining glory, raying down to us, sending its energizing rays to fill our body. Let us think of it, try to picture, try to feel the warmth within, especially the heart area, and then let it spread, this warmth, all through the body to fingertips, toes and top of the head; feel the glow, feel worthy of this beautiful energy, give thanks for it, for it is truly God-energy, empowering, rejuvenating and above all, healing.

Then let us give thanks for a brand-new day, and the chance to try again, once more to be more compassionate with everyone, but especially ourselves, so that we can accept ourselves for this day as we are, but trying once more to be a better person. And when you leave your bed, stretch your body, facing the Sun, or the Light, breathe it in, and give thanks for new opportunities, KNOW that good will come to you today, in beautiful unexpected ways; be open for lovely things, and above all, know in your heart that you are never alone—always there is your Guardian Angel by your side. You can be open to his/her energies, thoughts, just between yourselves.

Listen to the thoughts that flow through; reject those you don't need by pushing them away until they disappear into the mist, and then hear other thoughts that come unbidden—yes, aha! I could do THAT, something that had not occurred to me. THAT is listening and BEING the lovely person that you are. KNOW that you ARE a beautiful soul, striving like others to allow the best to come forth, and so letting out the JOY, and your heavenly Light. You ARE so loved!

The Orange Ray

Orange is a faster wavelength than Red and is the colour of calcium. Remember that Orange recharges the Etheric Body (Indigo heals it).

Orange brings JOY into our lives; it is a lovely energy and promotes:

- cheerfulness, confidence, compassion, enthusiasm;
- freedom from limitations; gives us new courage;
- independence, initiative, leadership, a person who can bridge and the one with optimism;

Orange also:

- gives tolerance, understanding and upliftment;
- helps with asthma or chest conditions;
- provides willpower to help initially to give up smoking; for this one needs to immerse oneself in Orange or Peach;
- raises the vibrations; like sunshine, is positive, sociable, warm-hearted;
- helps heal the reproductive area;
- brings one out of shock, relieves past hurts;
- links to intuition to help us master our own destiny;
- helps release grief;
- combines physical and mental energy;
- is a powerful tonic to all conditions of the nerves.

Orange brings out all creativity (especially Peach, which is Orange with added White) and is good for overcoming bad habits.

There is a need to breathe visualizing Orange light. We are trying to cleanse, purify, lift and then vibrate more quickly, linking to the stomach area with Orange. This is the belly-dancing area, where we can have fun moving to music of our choice. Orange is a stepping stone to lift us to the next chakras up the ladder—the Solar Plexus and the Heart.

We suggest NOT using Orange light or Orange oil higher than the lungs after 4 p.m.—it may be too energizing and prevent sleeping.

[If we lack vitality then we know the stomach energy is depleted.] The depletion of one area automatically affects the area next to it. A depletion of the reproductive area stops proper digestion and if, for example, there is over-stimulation of the energy above it, then the digestive centre is caught between them, not knowing whether it is coming or going.[52]

—Lilla Bek and Annie Wilson

Visit the author's website at

www.angelfire.com/ca/crystalbowl

for resources including colour and sound therapists, Chakra Oils, tuning forks, distributors of crystal bowls, and other useful addresses.

Chakra #2—Sacral

This Orange Chakra has a connection with abortion, which creates an energy blockage here; also, for women on the Pill, usually the Second Chakra is not functioning well.

This Chakra represents the Water Element, which brings the lesson of Divine Peace—not an easy path—making one intensely sensitive and receptive, needing the peace and quiet of one's own Sanctuary.

Orange stimulates the action of the Etheric Body; it removes congestion and increases the flow of prana.[53]

—Alice Bailey

AS THIS CHAKRA OPENS	WHEN IT IS FULLY OPEN
As the Red energy becomes tinged with the mental-emotional Yellow it becomes Orange, and as the centre opens and is therefore more balanced and flowing, there is a better understanding of reproduction and digestion, the working of the wonders of the body, fine-tuning the art of loving in the physical sense, therefore beginning to reach up to the Solar Plexus and the emotions, caring for the monthly cycle in the female (AND the male), moods, expression, warmth, being watchful about true balancing, having more care for diet for self, more towards loving one's own reproductive area and therefore, what one puts into the body.	Liver, spleen, eliminating organs all work in perfect harmony. The Creative centre is activated, beautiful New Age children emerge, they wrap each other in Light. With love, sexual intimacy stimulates, recharges and balances all chakras. Sexuality is the waterwheel of life!

Music for the Sacral Chakra

Within Pachelbel's Canon in D is the colour Orange (for the Sacral Chakra) and the musical note of D. This is the area of the Tan Tien, the bodymind connection that links physical and mental energies, so it seems natural that this particular piece of music is beneficial for insomnia as well as for mental clarity (has been found especially useful when studying).

Also:

- Mahler's Symphony No.1 in D Major
- Music by Maurice Ravel

Case Histories with Crystal Bowls

For the Sacral Chakra, we can play the Note D Crystal Bowl while toning the same note, alternating with the complementary Note A while visualizing Orange.

"I heard the D Bowl played, and saw the colour Green which changed to Red and Orange, then White Light—I felt in my body relaxation and wellbeing." —E.D.

Client had cold, toothache and sore lower back. We used Orange silk over body, toned together and played the D Bowl. Felt better. —M.S.

I have played the Note D Crystal Bowl (Orange) for an MS patient in a wheelchair. It was at a meeting of the MS Society—the Bowl was played at his feet with his permission. After a while he said he felt tingling in his legs (so did all the others in wheelchairs who were listening). How wonderful if this could be played to them several times a day. —R.B.

V.M., a nurse, had an infected colon and her energy was low. The Orange lamp was used with Orange oil on tummy. We toned together and the Bowls were played, especially Note D (Orange). She felt very peaceful, and later reported no more problems with the abdomen. —V.M.

Client had pain in left hip, leg and sciatic nerve. Used Orange oil plus D Bowl—leg was much better. Plus Orange/Yellow lamp on lower back with toning. Pain gone. —K.B.

Using Orange oil and Spectro-Chrome: After five weeks was able to give up smoking and had no more menstrual cramps. —J.D.

Orange lamp together with toning freed stiff neck and shoulders after one session. —C.M.

The Crystal Bowls were instrumental in the healing process of my ovarian cancer. I was diagnosed with this last year; the tumour was the size of a canteloupe. I had healing with Polarity Therapy plus the sounds of the Crystal Bowls notes C & G, which are complementary to each other and very powerful. When I had the surgery in January the doctor said I would be dead in six months, but on returning later for a checkup he found there was no more cancer! The energy and sound opened my chakra system and brought forth information and emotion to help clear and heal my body. I recommend any facilitator who knows and can play the Crystal Bowls as a wonderful find! —K.S.

Meditation: A New Day

Imagine yourself sitting on a balcony overlooking the sea, with mountains in the distance.
It is early morning . . . dark . . . but with a warmth . . .
and a silent hush everywhere . . . an air of expectancy.

The water is very calm, and as you gaze toward the distant mountains . . . snowcapped . . .
the Sun begins to rise in the eastern sky behind them—
first red, then changing to orange . . .
flooding the whole sky and reflecting in the water . . .
getting more golden and brighter, tinting the snowcaps. . . .

In this glorious light of a new day, you in-breathe with a sense of wonder . . .
excitement . . . for what the brand-new day will bring . . .
all the possibilities of challenge . . . re-creation . . . participation . . .
love for fellow Beings . . .
a great sense of vitally good health and awareness. . . .

Contemplate for awhile . . . with Joy in your heart.
Then, above all, give thanks for Being. . . .

Chapter Thirteen

Yellow— The Spirit of Wisdom

> The condition for clear and perfect reception is one of stillness of mind and silence, not only on the outer plane but deep, deep, deep within the Inner world, the Inner place. I am Divine Peace. . . .
>
> —Author unknown

On Walking in the Path of the Sun

Come, walk with me for awhile in the warmth and Light of the Sun—the spiritual Sun, for so it is, shining behind the Great Sun, invigorating, enfilling us with Light and Warmth and a feeling of goodness and wellbeing.

We tend, do we not, to take for granted all these comforts when we are enjoying the sunshine, but without it, we are desolate, depressed, full of woe, longing for the Spring. How can we walk within the Light of the Sun all year round wherever we are, and be healed?

Just think for a moment—let us use our imagination. Find a beautiful place where we love to be, just simply find ourselves there, taking in the sun's comforting rays, feeling so uplifted, nourished, cared for. Feel how grateful you are just to bask in its rays for awhile. This is your imagination, but if you take the time and relax, it can be very real; we can be actually there, experiencing in our mind's eye, feeling absolutely rejuvenated. That is the great power of the mind—the mind's eye sees it all.

This we can do on a daily basis, if we will but take the time. It doesn't take long, perhaps 10–15 minutes, but very precious time that is. For we desperately need the rays of the Sun, to enfill us, to rejuvenate, revitalize and to carry us forward during the day—the courage of Leo in the Orange rays, the Will too—so that we can dare to do what we need to do.

We so encourage you then, to take time to refuel yourself, to go on a flight of fancy which can become so real; and in dark days, retreat for a precious while to the sunswept horizons, bask in the sun's rays, breathe them into your Sun-starved body, mind and spirit—become whole again, entranced and beguiled, filled with the holy healing Yellow/Gold rays of the Sun—and be well, completely whole.

The Yellow Ray of Wisdom

The Yellow Ray is the Christ Ray—uplifting, the closest to Spirit. It has the most Light of all colours.

Just as the Sun is the centre of our Solar System, so the SOLAR PLEXUS is OUR centre—the second brain of the nervous system, the third magnetic colour, the doorway to the fourth chakra—inspiring us to reach higher vibrations and more into the Green Chakra of the heart.

People who watch the sunrise can expand their Solar Plexus with Yellow, so much that they become enormous. Yellow represents the love of knowledge, intellect and mental desire, and it helps to create self-confidence and courage. When GOLD is added it becomes Wisdom.

Yellow helps us to expand boundaries. It is not an easy colour to wear, but when we do, our clothing showers this vibration on ourselves and others. Yellow allows no security; it gives a wide-open feeling—you feel vulnerable, as if everyone knows you.

Scientists have Yellow in their aura, and people on the Yellow Ray—those for whom Yellow is the predominant colour in their aura—are bookworms. They love to study, but may get constipation sitting! Yellow is a very balancing energy, is cheering. One gets hunches, it makes one spontaneous. Yellow Ray people are on their own Pathway, making unique contributions to the world.

Yellow is the manifestation of the mental, creative vibration. Its use is beneficial in libraries and study rooms. The Sunshine School of York, England found that retarded children learned more readily in a yellow classroom.

The sunshine quality of this colour emphasizes the Joy of Living. Whenever we feel depressed, wearing Yellow will uplift our spirits. Too much Yellow, however, may give the impression that we are robots of reason and devoid of feeling. It may also make us appear too critical. When we are deficient in Yellow, perhaps raised by critical parents, we lack confidence and are confused. Not having been encouraged as children, there is the feeling that others control our life—therefore a need to bring Yellow into our Being. Sending Bright Yellow—the colour of daffodils—to the Solar Plexus Centre helps to clear blocks, develop understanding and relieve resentment.

Yellow is for thinkers—a thinker does not give in, he thinks things through. Yellow helps with clarity of thought, and a Yellow person loves to be admired. She/he enjoys freedom, Joy, laughter and fun and gets rid of pettiness; however, a fear of responsibility can hinder growth. A person out of touch with their personal power, thus unaware of their Solar Plexus area, can have digestive problems.

Yellow moves things along (natural laxatives contain the yellow vibration—for example senna root, prunes, golden figs and yellow fruits), and helps people to see good in all things, to be compassionate, to follow their intuition and to let in the sunshine. The complementary colour needed to balance Yellow is Violet.

NOTE: Yellow should never be used if one is suffering from acute inflammation, fever, delirium, diarrhoea, overexcitement or palpitations of the heart.

Chakra #3—Solar Plexus

The sympathetic nervous system is the organ of the subconscious mind. It is centred in the nerve mass located at the rear of the stomach known as the Solar Plexus—the brain of the nervous system. The vagus nerve, which connects the cerebral region to the thorax, branches to the heart and lungs before it passes through the diaphragm and links in with the sympathetic system.

The Solar Plexus is the Sun of the body. It is the central point of distribution for energy, which travels via the nerves to all parts of the body; from the body, the energy extends etherically into the aura and becomes living, colourful, radiating life energy and vitality.

Chakra #3 is the Power Chakra and its task is to overcome inertia till energy is produced easily—energy equals Power. When this centre is open and functioning well, there is a deeply fulfilled emotional life, but when it is closed there will be blocked feelings—maybe there is a block between the heart and sexuality. This is a very important centre for human connectedness.

The Third Chakra works with the Third Eye. Malfunctions are diabetes, hypoglycaemia, stomach problems and ulcers.

Yellow is the colour of the Sun and Yellow Ray people are stimulating, but they tend to ignore the body. The Yellow Chakra takes in more Mental qualities, felt as the first level of intellect. These people prefer more sophisticated and classical types of music.

This chakra regulates and distributes metabolic energy throughout the body as well as the combustion of food into energy. The digestive system is an important part of the process, as the stomach coordinates with the liver, gall bladder, pancreas, duodenum and intestines, substances moving along the intestinal tract. All is a rhythmic process, including breathing and heartbeat.

. . . The solar plexus is the reservoir of our vital forces, the accumulator of all our energies, and if you know how to refill it every day you will have a source at hand from which to draw the energies you need at any moment. . . .

In the ancient treatises on alchemy there is constant reference to a kind of oil or essence that possessed marvellous properties: it could give health, intelligence, beauty, knowledge and so on. In point of fact, all living things, plants, animals and men, can distill this essence. It has been variously called the true sap, prana, the elixir of eternal life, etc. Some people call it magnetism.

. . . [T]his essence can be found everywhere. Plants draw it from the soil, the air and the sun's rays and, thanks to it they produce their sap. The sap of plants is a symbol of the living sap that flows in us. And where is the spring from which it flows? In the solar plexus.[54]

—Omraam Mikhaël Aïvanhov

In assessing the health of this centre, tight hard stomachs, large pot bellies and sunken diaphragms all indicate excess or deficiency:

- Large bellies: possibly large ego force for power to dominate, control, and a need to be noticed;
- Weak, sunken diaphragm: fear of taking power, withdrawal into self, fear of standing out;
- Excess weight in general: can be a third-chakra malfunction, the body not properly metabolizing food into energy.

Most of us are out of balance—everyone has depleted energy in one centre or another. If the mind is fuzzy and it is difficult to concentrate, the Solar Plexus is out of balance. A mind overactive with worry affects the Solar Plexus; the stomach will also suffer.

When we control emotions and anger, the Solar Plexus shuts down. Children often have stomachaches and shut down at the age of two. The best thing is to get them to lie down and palpate the Solar Plexus.

Research on the Solar Plexus shows that:

- *schizophrenics have a very bright, overcharged solar disc;*
- *in shy people, the Solar Plexus can overcharge;*
- *those who act or do public speaking recharge the Solar Plexus; they release a glow which exhilarates the whole audience;*
- *we can control our energies by balancing the lower and higher self;*
- *in a very relaxed body, we can be in touch with our Higher level;*
- *linking with the colour Gold (wisdom) helps us to link with the God-power and the higher aspects of creativity.*

AS THIS CHAKRA OPENS

On opening there will be a greater feeling of Joy and a feeling of expansion in the diaphragm area. Emotions can be released immediately instead of being held bottled up. Assimilation of food will be more complete and therefore more nourishment will be gained. Because this centre is connected with the mind, that too will be clearer and it will then positively affect the heart centre—touching it, like an encouragement.

WHEN IT IS FULLY OPEN

Bodily health is taken care of; emotions will be integrated.

All three lower chakras are involved in the opening of the Heart Centre, which is the level of world service and commitment. Once the three lower chakras are opened, they begin to operate in service to the three higher ones: the Throat, Third Eye and Crown Centres.

Yellow, Red, Blue or Green Filters for Dyslexia

Lucy Hawking, writing in the *Daily Mail*, reports the following:

> Dyslexia, or word blindness, affects 2.5 million people in the U.K., but until recently scientists have not been able to find an effective treatment, let alone a cure, for this distressing condition.
>
> Research has shown that people with dyslexia are more likely than others to suffer from light sensitivity. Dr. Wilkins believes these problems can be caused by strong black writing on a white page, causing an overload in the part of the brain which deals with sight.
>
> It was ten years ago [actually in 1988] that American psychologist Helen Irlen first hit upon the idea of using coloured lenses worn in glasses, to help counteract the reading difficulties experienced by dyslexics. She covered text with a variety of coloured sheets of see-through plastic to see if a particular colour would make reading easier. Her findings, though, were largely dismissed by the medical community as "unscientific".
>
> Scientists at the Applied Psychology unit in Cambridge (U.K.), however, have now proved that spectacles fitted with coloured lenses can ease reading and writing difficulties. Although scientists have recently isolated the gene they think causes dyslexia, there is no known cure for the condition, nor any proven treatment. . . . Coloured lenses seem to help block out this interference.
>
> The lenses work in a similar way to looking at the sea on a bright day through Polaroid sunglasses: the "glare" from the page is cut down so outlines of individual letters appear sharper. But Dr. Wilkins and his team wanted to find out why this happened. To help study the effects of coloured lenses, Dr. Wilkins developed a machine known as the Intuitive Colorimeter. It had already been observed that the colour of lenses required differs greatly from person to person, but discovering which colour worked best was a process of trial and error. The Intuitive Colorimeter can pinpoint the tint required far more quickly by projecting colours onto a page. The colour and depth of the light can be adjusted until the child can read without distortion. The test usually takes less than 15 minutes and once the right colour has been found, lenses are made up in this tint which are then worn as spectacles. Eighty per cent of volunteers in Dr. Wilkins' study reported continued improvements in reading abilities after a year of using tinted lenses.[55]

In the Vancouver (B.C.) area, there are special teachers using four colours of filters over a page. Eva Reycraft states: "We've had tremendous success with these for reading. More sensory integration is often needed, but these work for most."

CONTACT:
Eva Reycraft & Associates, Vancouver, B.C.
Tel (604) 266-1981; Fax (604) 261-1976

The Panic Button and Fear

The sun is shining, you are in a wonderful space, feeling good, loving every moment. Nature is beautiful and so are those around, when Wham!—suddenly the news is bad. Something happens to spoil everything; you are in a state of frenzy. What to do? What will happen? All sorts of crazy thoughts rush through your mind and your body reacts. You feel horrible, especially in the solar plexus. Adrenaline rushes in, but still you feel rather helpless; you have to make a big effort to pull the threads together and take appropriate action.

Whatever happens to each of us to put us into a panic state is basically fear—fear of the unknown, fear of the future, fear of a sickness. Many different fears all come together and press the panic button.

What is fear? Where does it come from? Where do we feel it in our physical bodies? In the stomach, the solar plexus, in our heart? It is a feeling that comes through a not-knowing, a misunderstanding of how we can be, an irrational depth of knowing, a complete understatement or distortion of the Truth.

When we harp on this misconception continually, it builds into something fierce, usually completely out of all proportion—a lack of proper knowing; whereas if we sat down and in the quietness allowed ourselves, without judgment, to think through the whole idea clearly, then would we be free of all judgments and feel a great calm pervading all our bodies.

What can we do to avoid fear, to rise above it, stay with that inner Peace ALL the time? There are ways, but firstly it must be to find the stillness within, to actually stand still for a moment, to centre oneself with the breath, simply listening to the breath rising and falling, very calming. Then to link to one's Higher Self and ask for direction—to be guided as one progresses in the daily task, to be shown the next best step—and in this initial calmness the link is made with your Higher Self, the Shining Ones, when much magic can occur. Ideas will come swiftly, to be acted upon calmly (above all, keep yourself calm as possible). Do the important things first and all the rest will fall into place. Keep affirming "ALL IS WELL—I AM CALM WITHIN" and breathe deeply, breathe in the Light. Fill your whole Being with the Light.

Thank you, God, for showing me so clearly my pathway now, what I must do first! And having done that, everything falls into its rightful place. No more "flight or fight," which does enormous damage to us—leave it all behind. Forgive and forget, and go forward feeling absolutely free, uplifted, ready for the next test, being fortified, and loving ourselves for being so free.

We need to practise this daily so as to be ready for any occasion. The Light, the spiritual Sunlight, is really all that matters—is the importance in your life on Earth. All else fades into insignificance when we put the Light first. Try it—fill your Being with it. Walk in the Light. Be the Light Centre that you truly are.

Decrystallizing Fear

Take long deep breaths . . . slowly in and out . . . drawing in Light with each breath . . . exhaling thoughts and agitation. . . . Ask the body where it is holding fear (simply ask the question). . . . Continue deep breathing. . . . Notice where/when you feel sensation in some part of the body . . . or hear the answer. . . . Trust . . . go to the right place . . . rest there . . . feel, connect deeply in the area. . . .

How does fear look, smell, feel? What colours do you see there? Ask the body what colour it needs to dispel/dissolve fear. . . . Accept the first colour you see or feel. . . . Draw it in, and keep on doing so until your body can receive no more. . . . Be aware that the colour is absorbing and dissolving all fear within. . . . When you feel the change, open your eyes.

When you use colour to dissolve the dread, the cause is actually dissolved too. When we change the energy, we change the reality.

Lemon—Visualization

In your mind's eye, place yourself in consciousness in your pituitary gland. In the right lobe see yellow. In the left lobe see green. Slowly merge these two colours into the centre, as lemon or lime. Feel the perfect balancing of the right and left lobes. In this balancing, feel all the organs of your body, all the muscles and bone structure saturated with the lemon colour, starting at your head. Feel it flowing down your neck, shoulders and arms, through your chest and abdomen, through your thighs, sacrum and base centre and then through your legs and feet. Allow the deep cleansing of this lemon colour to help you release all unwanted matter. Allow this lemon colour to perfect the processes of nutrition and repair in your whole system. Allow it to balance your right and left brain functions.[56]

—Audrey Ann Lowrie

Tinted Lenses Create Problem

What of tinted glasses that are so freely prescribed, or on display in the optometrist's office so that people can be attracted to them for aesthetic reasons?

At one point, a doctor referred his patient to Dr. Jacob Liberman to find out why she felt ill all the time. Dr. Liberman interviewed her, then asked her to remove her spectacles, which he examined. They were pink-tinted. He told her to leave them with him, and sent her home for a week with another (clear) pair.

A week later she returned, feeling perfectly well. It meant that she already had sufficient pink energy in her body. The lenses only added to it to make too much pink, and put her out of balance!

So what colour lenses are you wearing? Do you really need extra colour? The same applies to wearers of contact lenses.

Pain in the knees may be the fear of going forward!

The Clear Yellow Beam

The Light of the Yellow Beam comes directly from the spiritual Sunlight. One can picture it as a spiritual replica of the Sun, almost in the same way that the physical body has an etheric counterpart, an exact replica but of a higher, Light vibration, not always seen by everybody.

Picture then this beautiful duplicate of the Sun, in all its glory—a great golden Yellow beam—raying down upon you. Its rays are warm, comforting, uplifting, penetrating and very healing. They enfold you and also top you up, make you whole, quieten you, softly relax you, touch you where you need to be touched.

Continue to visualize this beautiful spiritual Sun in your meditative state, pouring down upon and into your very Being—Light and Love filling every cell of your body as you breathe it in. Feel it moving through your body to the very tips of your toes, fingers, nose, ears, eyes—wherever you need it. Feel that wellness pouring through you. Be happy, be well and be at Peace. . . .

When mankind feels free and lighthearted, paler colours are worn.[57]
—Marie-Louise Lacy

Studies at UCLA have shown that when subjects gaze at warm hues (Red, Orange and Yellow):
- *blood pressure rises;*
- *brainwave activity increases;*
- *respiration is faster.*

Music for the Solar Plexus Chakra

The Solar Plexus Chakra, the chakra of empower-ment, resonates in the key of E. The Note E stimulates the intellectual activity of all life forms, but humans in particular. Research has shown that Mozart's music is especially effective for analytical and mathematical thinking. His Symphony #39 in E♭ Major stimulates the Solar Plexus Centre. Mozart, who could express the gentler side of music (he is said to have incarnated on the Pink Ray of unconditional Love), has been hailed as "the greatest of all musicians."

Case Histories with Crystal Bowls

Play the Note E bowl and tone the Note E for this chakra, while visualizing Yellow/Gold enfolding you.

The 18" Bowl (Yellow/Note E) affected me very deeply and profoundly around the Solar Plexus, clearing and balancing without any discomfort. I loved it! —A.G.

MEDITATION: GOLDEN ENERGY

As you sit relaxed, but with spine erect, picture the beautiful YOU out on a lawn. . . .
It is a perfect day . . . blue sky overhead . . . Sun shining warm on your body. . . .
See yourself standing tall, raising your arms and face
as you begin to breathe IN slowly. . . .

Feel yourself reaching Heavenwards towards the Sun, breathing in the glorious Sun energy
from above into the Crown of your head. . . . HOLD for a moment . . .
giving deep thanks. . . . Then, exhaling very slowly,
see this Golden Energy permeating every cell of your body . . . healing. . . .

Lower your arms and slowly bend forward, picturing the Golden Light going down your legs,
through your feet and into the very Earth itself . . . nourishing Mother Nature . . .
the soil . . . the roots. . . . Become a tree . . . HOLD for a moment. . . .
Then, as you commence again to inhale,
see the beautiful Golden Light of Earth energy rising within you.
Bring your hands together, and with the Light coming up your legs, trunk,
arms lifted skyward in blessing, HOLD . . .
till you want to lower your arms wide,
sending the Golden Energy into the Universe as LOVE . . .
and PEACE. . . .

Breathe yourself slowly back to "Earth" and open your eyes. . . .

Chapter Fourteen

Green—
Ray of Balance & Harmony

There is a secret core of meaning waiting to be discovered in every life, and those who find it are granted not only greater wisdom, but also greater joy in every day and in everything they do.

—Deepak Chopra, M.D.

Your Inner Healer

We are here to help you. We ask you to be patient. We ask you to relax. We ask you to Love, Love, Love. Go into the garden and consciously draw in the beautiful energies. Fill your whole body with Light from the Sun each day. Walk with JOY. Your throat needs expression of Joy and Love ALWAYS. So in your Heart be at Peace and release all anxieties, frustrations and disbelief. Just be quite natural. Listen above all to what you have been told. Gradually connect with your Inner Healer wherever you are. Then know the next step to take.

Feel assurance you are on your healing path and recovering. Your Inner Healer is your companion. Speak as with him/her whenever you have a question. Above all—LISTEN—LISTEN—LISTEN. Become conscious of the Light more than ever before. See yourself enfolded in it—like a cloud. See yourself doing things you hope to do without any anxiety—only with Love and Light. When your body is filled with Light there is no other room for anything else—just JOY!

The Green Ray

The heart is the seat of the Soul. . . . The only way to purify the heart is to lose our over-riding fear of being left alone. We must understand the myth about security. . . . Only then will the heart be ready to merge with a higher consciousness. . . .[58]
—Lilla Bek and Annie Wilson

Green is in the middle of the spectrum and stands for neutrality. It is Nature's colour of renewal, regeneration, rejuvenation. Green is the balancer and harmonizer of the Universe. It links to success, prosperity, hope and eternal youth, and is excellent for children and animals.

If we are lacking balance, it seems that we are pushed into situations or relationships where we are forced to introduce the colour Green. Doctors and nurses and those in the caring professions are often associated with the Green Ray; it keeps them calm and balanced. People with Green Fingers emit lovely energy, but may give too much, to their detriment.

Olive Green denotes deceit and treachery; murky Green spells envy. Darker Greens and the green of envy we can dissolve with Orange—otherwise it is a wasted energy. Using the Emerald Ray through the Brow (Third Eye) Centre can help. Emerald is very rejuvenating—it holds a very high vibration, raises self-esteem.

Green is the restorer of peace; it helps balance the etheric body. It is a balancer, a relaxant and is restful and restorative; it allows oxygen to permeate every tissue. For emotional upheavals or bottled-up emotions, go into Nature—they will release and relax, thus preventing palpitations.

Green is excellent for anger, high blood pressure, headaches, the heart; it has an equalizing effect. It acts as a tranquillizer. Green helps one relax and sleep. It also helps expel toxic wastes and viruses.

Green is great for all nervous disorders, for the nervous system and breakdown. Wear an Emerald Green scarf, or put one on the back of the chair and rest your head on it. Go into Nature, breathe in and out—especially where there are fir, eucalyptus and cypress trees.

Green and Pink go together—imagine being loved and spiritually caressed and wanted! Use Green on the pituitary to bring balance.

When the heart area is open, then compassion, empathy, understanding and healing come. When the heart hurts, take down walls, let out the grief. We all need more Green. Many today have heart troubles and have not been able to link with Soul—Green relaxes it.

Chakra #4—Heart

Location: First Dorsal—mid-chest
Function: Love/compassion
Organ: The Heart. This chakra has a profound effect on the lungs, heart, blood, vagus nerve (which supplies the heart and lungs) and circulatory system.

The Age of the Fourth Chakra is one of peace and balance between Spirit and Matter. This is the central point of the chakra system, also the centre of our Essence, the seat of Love, the Mirror of the Soul. Love is the pure essence of true healing.

The Heart has a specific mission: to keep alive the whole system—to integrate, balance the body and mind, to bring a radiant sense of wholeness to the entire organism.

As we move up the spectrum, we meet people who are beginning to be aware of inner silence. Green people are full of outflowing Love, and in return will receive Love from others. All love affairs begin by overcharging on the Green, but in a beautiful way. Green people make good gardeners and farmers.

A Green person ranges into the Blue Ray, is drawn to water, lakes and the sea. Temperamentally more balanced than other colours, Green Ray people can also make excuses for other people, have a soft heart and be easily hurt.

In relationship with Self, the heart is the centre, the balancer. It balances lower and higher centres because the main energy pathway in the body is a current through the spinal column. Each chakra receives a charge by being in alignment with it, and the Heart suffers the greatest loss if "out." Imbalance here can throw the entire system off balance. Love is a state of being in harmony with self.

Imbalances in the Heart lead to:
- blood diseases—leukaemia;
- disorders in thymus and immune system;
- all types of heart dis-ease;
- inability to balance emotions;
- emotional extremes of all types.

When the Heart is full of pain, the basement (subconscious) must be cleared, freed, looked at, examined and loved. The Heart must be open, clear and shining. When the Heart opens, the Throat opens. If the Heart remains closed, so too do the Throat and Base Chakras.

So we must work on emotions, get the Heart clear—then the cell structure changes and Light enters. The more refined the Heart, the more the smallest misuse will leave a trace. The use of stimulants as well as emotional upheavals will leave a scar. Those who smoke create rings over the etheric body in that area!

Be discerning about teachers—ask yourself, is this person of the Light? The present age is the time of small groups, committed people; here there is tremendous energy. The New Age is of the Cosmic Heart. Listen to your Higher Mind in order to adapt to quickening energies of the Earth. Each time we channel healing, we partake of small doses of the Cosmic Heart.

The lessons of brotherhood come through communication and understanding of the needs of others—through balance of Heart and mind.

AS THIS CHAKRA OPENS

– one can have far greater compassion;
– there will be ease in breathing, the ability to take fuller breaths—right up to the throat;
– one has greater understanding for others; a gentleness;
– one has a radiance; loving acceptance of oneself—then comes the same for others;
– one develops mastery of the immune system and the thymus;
– the Throat Centre opens more easily, followed by the Third Eye;
– our energy begins to flow beautifully.

The heart . . . responds quickly on a physical level to almost every situation . . . our adrenalin levels rise, and with constant stimulation to stress or "danger" our adrenals must suffer.

Today we hold in our emotional problems. . . . The heart, switched on and off like a light bulb by constant stimulation, will reach confusion point and eventually the bulb will explode.[59]
—Lilla Bek and Annie Wilson

WHEN IT IS FULLY OPEN

– we feel a lightness;
– we are refreshed from the etheric level;
– we are fully in charge of the body—not aging, only maturing;
– we are in complete communion with our soul;
– we are very creative—everything we touch flourishes;
– we will be in total balance;
– we think of something and it happens!

When the Heart is totally open, one commits oneself deeply to service to others. Examples of this are Mother Teresa and Paramahansa Yogananda.

Any illness is a direct message to you that tells you how you have not been loving who you are, cherishing yourself in order to be who you are. This is the basis of all healing.[60]
—Barbara Ann Brennan, "Heyoan's Meditation on Self Healing"

Love and Colour

Love is Pink unfolding. Pink is unconditional Love—for self first, then for the whole world. Love is empathy, compassion. It is feeling for others, it is nurturing, it is the finding of oneself in the space of one's Heart, to love oneself in the understanding of being all Love, of fulfilment, of being able to tune in for the highest needs, not only for oneself, but for another, with an all-encompassing knowing . . . knowing what is best for the moment . . . and leaving the rest in Higher Hands with full confidence.

When there is Pink, there is Love. Use Pink clouds to encompass people who are out of sorts, who cannot communicate. Let the Love enfill their hearts—use Pink.

Forgiveness Prayer

The following is an example of a Forgiveness Prayer. It is one that I have used for a long time.

All that has offended me, I forgive. Whatever has made me bitter, resentful, unhappy, I forgive. Things past, things present, things future, I forgive.

I forgive everything and everybody who can possibly need forgiveness in my past and present. I forgive positively everyone—I am free and they are free. All things between us are cleared up, now and forever.

Music for the Heart Chakra

- Rhapsody in Blue (Gershwin)
- Strauss Waltzes
- Tibetan Bells
- Water Music (Handel)
- Blue Danube Waltz (Strauss) has same rhythm as the heart—and regulates it.
- Pieces by Debussy, including Clair de Lune
- *Hari Om Mantra for Meditation* (flute), by Ann Cunningham; tape produced by Yasodhara Ashram, Box 9, Kootenay Bay, BC V0B 1X0.
- *Hari Om Mantra for Meditation* by Swami Sivananda Radha, Timeless Books, Box 50905, Palo Alto, CA.
- *Celestial Light*, composed by Hildegard von Bingen (1098–1179), words and music written in open form, chants sung by Tapestry, CD-80456, Telarc Digital, 1997.

The pale Green of Spring stands for lightness, especially in the Heart. Think of dancing in a garden. Do you remember the Green of a white hyacinth before it blooms . . . or perhaps the bright Green of lime jelly?

A Strong Heart Centre

Imagine for a moment a gossamer-like thread weaving its way around and around a filmy pattern. This weaving goes from the inner core to the outer edge, slowly forming in beautiful intricacy, with colours appropriate for the time, for the moment. And into this beautiful pattern are woven most delicate hues, of kindliness and understanding, Love and compassion (Green), Peace (Blue), Joy (Orange), devotion (palest Lilac), healing and revelation (Indigo), and the deep splash of Red for courage, forbearance, strength and mobility.

All these and many other tints and shades—for whatever is for the moment in time—all swirl and intertwine, making all the weavings within our Heart Centre scintillate . . . glow with a warmth that we can actually experience.

It is only when we harden our Heart for various reasons that the tints turn to darker shades, and we feel an icy coolness around us. Then do we have to work harder on ourselves, to figure out where we need to change in our attitudes, where we need more understanding. Above all, we need to let out the Joy, be as the Sun once more, shine brilliantly, with all the colours scintillating, weaving in and out, touching each other, commingling, and THEN radiating out to others—for first must we nourish ourselves.

In your daily round then, remember this: fill your Heart Centre with colours. See them, feel them and then allow them to shine forth, reflected in your eyes, face and skin—shine forth and be a Light for all to see and feel . . . THAT is happiness!

Case Histories with Crystal Bowls

I heard all the Bowls plus Heart F note. The day after playing, in my bones and in my Heart area, it felt resonating and clearing, then all discomforts eased. —L.K.

K.F. had many migraine headaches since childhood—more during menopause. Pain felt down left neck, shoulder, left hip. Used Green aromatherapy oil for pain—played Bowls, especially F, and toned together. All neck pain went and headache too.

Ten-inch F Bowl was played—it was very powerful while at my feet. I felt spiralling vibrations throughout my body; it was very energizing. —A.G.

Friendship

How very lovely, very special, to have a friend who walks with you through the many phases of your life—who accepts you as you are, as some say "warts and all"—and understands your true feelings, is there with a kind, special word just when it is needed, who offers a shoulder to cry upon, who also has a tremendous sense of humour (my friend and I call it a "sensa" and know what we mean!). How wonderful to see the funny side in everything, to laugh at ourselves, our foibles and mistakes—and it matters not that she/he is at the other end of the Earth, the ties are there through thick and thin, through special moments, the sharing of Joy and the sadness, the delights and the sore spots.

Then there is the special encouragement in whatever you are doing—new projects, your children. Sometimes you both seem to run the same pattern! A very special understanding then, and even if there is a space, for whatever reason, when there is no communication, bet your life it will return with a bounce. You carry on as if you know exactly how she feels, and yes, it's true, spot on—intuition? Telepathy?

Treasure this friendship then. Realize how very honoured are you both, how very special, and because of your closeness and understanding, your very open heart full of Love, then the world keeps turning. You know you are loved, and can return that Love twofold, with great thankfulness always.

Companionship

In your life, have you ever thought about, or dreamt about, a beautiful close companionship and all it entails? Have you dreamed of all the possibilities? Have you even pictured it in your mind's eye—the sheer beauty of it, which can be felt, but perhaps not seen? Do you long for this? Have you seen it in others? Felt the look in the eyes of your special companion? Known the nearness, but especially appreciated the understanding between two people? No need for voicing one's feelings, thoughts, but instead they are actually deeply understood in the silence between?

There can be a most beautiful ambience around two people who have such enormous understanding of each other that words are no longer always necessary. And beneath it all is a great Love, a spiritual Love and understanding, as if they had always walked together; and if you watch closely these dear people, you will see they are enfolded in a great Light—a warmth. It shines in their faces, from their eyes, and you can feel it touching you.

This companionship is a nurturing and an understanding from a deep level. It is as if they had always known each other and had come to Earth to be together and continue the nurturing—so there is an "un"earthly quality to it, and we need to rejoice in its happening, especially if it touches us.

How can I nurture this quality, this touch, this depth? Let me grow in its Light and realize that we have been so filled by our Guardian Angel souls—it is a SOUL quality. Let us be so grateful, acknowledge it, revere it, encourage it and be very joyful, for so it is.

We give enormous thanks for just Being.

Holy Matrimony

The Holy of Holies, the point of departure onto a brand new Pathway, Holy Matrimony is the entering in to another's secret Temple—not completely, but certainly a foot on the threshold of knowingness. It is like lighting a fire within the Heart of another, using your flame. This really means guarding and nurturing that fire from then onwards, to use every ounce of Love possible to do the nurturing, but with the understanding that it is God's Love which fills the Heart. And if you will it, this never dries up, for as we give so we receive, and in the giving it is replenished.

This fire in the Heart is a great energy. It fills you, overflows, catches your breath, sets on fire the flames within the Heart of your much-loved one, and after the initial roaring flames of passion, possession and nearness, there come the lovely steady embers, which need caring, loving hands, to make sure the fire never goes out, but holds very steadily. There may be flames for awhile while the embers are stoked and replenished—sometimes a crackling, sometimes a blue flame. The crackling we need too, for life is not meant to be smooth, and after the crackling, the blue flame of healing and understanding with compassion.

The fires of youth do die down, but where there is a deep Love and understanding with appreciation and gratitude, then the flames are still there, but gentler and wider. Then can one journey onwards together and the Holy of Holies, that space within the Heart Centre, becomes larger and all-encompassing. Others are nurtured too. The Heart is deeply comfortable and filled with a quiet Joy, knowing that despite all challenges, welcomed or unbidden perhaps, there is always this quiet centre, this Holy of Holies, where two can gather and give Love to each other, give support and also give thanks for the balancing, true understanding, and for the God-like energies which permeate the bond between.

This is Holy Matrimony—walking together in the God Light which radiates through both together and is felt by all who touch them.

Down Memory Lane

Have you ever wandered magically down this gossamer lane? All sorts of dreams and special people appear in your mind's eye. Your imagination soars into all kinds of possibilities and you become very engrossed on your journey through the past. You see people again that you haven't seen for ages, perhaps since they transformed to the next plane. You see them now with more mature eyes, with more understanding, and if you are taken into the darker side of happenings, now in hindsight you can see what happened for its true worth, without bitterness or resentment, but hopefully with more understanding—and it could be with laughter! Allowing your keen sense of humour to come to the fore—without blame simply seeing the occasion for the challenge it gave to you at the time.

These jaunts down Memory Lane now and then are so good for cleansing ourselves of what we feel to be outrageous injustices and infringements on our worth. Seen in the clear light of maturity, we can allow that the person infringing knew no better at the time, and within our hearts we can forgive, then move on, freed of resentment or guilt and misunderstanding. The really darker sides of memory such as abuse and incest can then be dealt with by competent therapists—we can ask to be led to the right one to help us.

Now is the time to cleanse ourselves of all that we need no longer—old memories from the "dis-used shafts" in our mine of the mind—to bring them up into the Light to see their worth, and have them dissolved with Love, understanding and laughter. Then when each of us is cleared of the past, but still retaining very happy memories which evoke Joy within our hearts, we can go forward in our lives with more Love and compassion for ourselves and others. We can begin to enjoy every blessed moment of every lovely day, not wasting a moment on regrets or what might have been, but really beginning to live in the "now," and enjoy life. Let this feeling become infectious and people will love to be around you.

Be at Peace then within your Heart, and let the Light shine through your eyes, from the very soul of you.

Meditation: The Rose of Your Heart

Within the stillness of your Heart Centre, picture a most beautiful Rose—
all silky, palest Pink . . . smooth to touch . . . softly curled petals . . .
half open . . . sweet perfume . . . beautiful to behold. . . .

Hold the stem gently and gaze upon this, savouring the gentle fragrance,
allowing it to permeate your Being . . . healing you . . . for aromas can heal . . .
can help to restore balance. . . .

And as you continue to gaze upon the wonder of its creation . . .
watch the gentle unfolding, opening of its delicate petals . . .
the baring of its Heart Centre to yours. . . .

Feel in the exquisite beauty of Nature's creation
the flow of wonder and love and gratitude to your heart—
that is your spiritual unfolding.

You too can be like the beautiful rose—exquisite, soft, gentle,
spreading a quiet radiance around you just by Being . . . a part of Mother Nature . . .
giving of your very own quality to uplift, enchant and heal. . . .
Savour the idea in the stillness of your heart. . . .

Chapter Fifteen

Turquoise— Ray of Tranquillity

> The witch doctor succeeds for the same reason all the rest of us succeed. Each patient carries his own doctor inside him. . . . We are at our best when we give the doctor who resides within each patient a chance to go to work.
>
> —Albert Schweitzer

On Being Gentle and Kind in Whatever You Do

Realize that gentleness and kindliness are the hallmarks, if you will, of the Initiate, that whatever is said and done is with this gentleness and being kindly. It must absolutely saturate our Being, ooze from us as it were, without conscious effort, so that the person on the receiving end feels cared for, feels understood, feels worthy of your kind attention—so much so that they would emulate you, that they could struggle to be like you in that way.

Gentleness comes from the Heart Centre. It is a need to make the other person feel good within, feel worthy, feel that nothing is hopeless, feel that after all, there IS goodness in the world, that someone cares. This melts everything within their hearts, makes them appreciate not only your way, but also themselves, that there are surely some possibilities, if they would only try, put their hand to their wheel.

What then of kindliness? This too comes from the heart, but also through the Blue Throat Centre. The expression, the nuance, the inflection of the voice are all there to express feelings, to make the other person feel they are wanted, appreciated, they are worthy of attention—and above all, they are understood for where they are, at this exact moment in time, with all the earthly experiences they have accumulated behind them.

Kindliness and gentleness cost nothing—perhaps a little extra effort and time—but nevertheless, they are the first steps towards initiation within expression, a part of counselling.

Let us then be more aware of ourselves and how we can best use these wonderful "tools of our trade"—to encourage, to send people on their way rejoicing that at last they are understood, and in their understanding they are truly armed to cope again on a brand-new day. And perhaps begin a cycle—that they can in turn be that too, and encourage others along their Pathway of Light.

The Turquoise Ray

Turquoise works with the Thymus Chakra. It is situated between the Heart (Green) and the Throat (Blue)—Turquoise therefore combines Green and Blue! It is a wonderfully cool, relaxing and refreshing colour that aids any inflammatory condition such as swellings, cuts, bruises and burns.

Turquoise is especially good for:

- Skin problems, acne, eczema, psoriasis;
- Stress, tensions—calms and relaxes;
- Removing toxic wastes and congestion from the body;
- Helping the immune system protect against invasion of harmful bacteria and viruses;
- Colitis, dysentery and fevers;
- Assisting in the elimination process;
- Helping clear sinuses, mental fatigue and hay fever;
- AIDS sufferers, especially in early stages.

Wearing Turquoise also helps with expression and keeping calm while speaking before groups of people. Turquoise is not recommended for people with sluggish or underactive conditions.

Turquoise is often used as a skin tonic—it tightens and strengthens the muscles, and rebuilds burned skin. Dinshah tonated his skin with Turquoise for twenty minutes every day, and he is reported to have had a flawless complexion!

Turquoise is a tranquillizer because it is cooling and relaxing. Headaches can be alleviated because of its relaxing qualities.

Stanley Burroughs suggested that a quick refreshing nap under Turquoise light after vigorous work or exercise can restore vitality quickly.

Janice Ellicott of London, England, on reading my article in the *International Journal of Alternative & Complementary Medicine* about the newly awakened Thymus Chakra, wrote:

> For many years I was puzzled why people felt such pain on the lymph reflex between the first and second toes, even though their other lymph reflexes were not painful and from their general health their lymph system seemed to be alright. After reading what little information there is on the thymus gland I concluded that the pain stretching down beyond the lymph reflex could only be the thymus gland. . . . Much more attention should be given to it on a physical and mental consideration of the patient. . . . As it is closely linked to the muscles in the body it should be very carefully and consistently treated for muscular dystrophy, MS, paralysis after a stroke and other disorders such as constipation, incontinence and persistent stiff neck. . . . I am now wondering whether thyme essence should be used on the feet when the thymus reflex is painful.[61]

Chakra #5—Thymus—The Higher Heart

Location: At sternum above the Heart Centre
Colour: Turquoise
Gland: Thymus

This is the centre where art is born. Pounding on your chest (like Tarzan did!) will stimulate the thymus gland into action. The thymus gland regulates growth and the immune system.

Thymus Practice:

- drink thyme tea—stimulating for thymus;
- love yourself in many ways;
- wear Turquoise gems, stones—chrysocolla and turquoise;
- breathe in Turquoise; visualize Turquoise into the thymus area;
- Use fish posture (Yoga) as chest expander, or do any stretching exercises for chest area;
- Listen to calming music—let it wash over your body. Sound is felt through every pore in the body; it helps to release tension;

Remember: All sounds touch the Thymus Chakra.

(For more information on the Thymus Chakra, see *THTCB*, pages 81–82.)

Turquoise—Visualization
See the colour turquoise enveloping you in a soft, hazy light. Breathe in this light. Allow it to be absorbed by the etheric webbing of the atomic body, processing it and sending it out to all the capillaries in your skin. See it beautifying and toning up your muscular structure and tightening up your flesh. Feel the gratitude of your face and body for this refreshing colour bath.[62]
—Audrey Ann Lowrie

Case Histories with Crystal Bowls

Sound the Bowl and tone with Note F# or any higher note, visualizing Turquoise. (It is a lovely idea to put inside the Bowl a note containing your birthdate. Playing the Bowl with this note in it may help guide the attunement to your Soul vibration.)

This person feels utterly drained—several deaths in the family over the past few months. We played the Turquoise F# Bowl near the Solar Plexus and Root, and at the feet. We used a Magenta lamp and oil on the Thymus. Feels very good, relaxed—and at peace. —P.A.

Has been very stressed for three years—spleen really weak; trying to regain energy. Used Turquoise oil on the heart; played Bowls including F#. Feels relaxed and very peaceful. —L.T.

Affirmations for the Thymus Chakra

I AM opening myself to the purity of the new.
I AM becoming all that I can be for the purpose of helping others in this changing time.
I AM keeping my words wholly loving and full of heart.
I AM expressing myself as a Child of God, and the Divine Will is always in control.
I CLAIM the power of this new energy centre for good.
I AM pure Love expressed.

Turquoise is imaginative, youthful, sparkling fresh, sensitive, clean, transformational, elevating.

The Turquoise Ray brings Peace, calms the mind and emotions—I AM in command of the situation.

Disorders in the thymus and the immune system can be the result of imbalances in the Heart Centre. And constant exposure to the following weakens the immune system:

- adulterated food;
- radiation;
- doctor-prescribed medications;
- environmental pollutants;
- stress, which aggravates the adrenals.

AFFIRM: "My immune system is very special—it works with great efficiency, totally empowering my body to be at maximum efficiency and wellness. I am completely balanced."

Music for the Thymus Chakra

- *Ancient Echoes* by Steven Halpern/ Georgia Kelly
- *You are the Ocean* by Schawkie Roth (zither and flutes), 1979, Heavenly Music HM-0103.

I have been given its purpose as an "I AM" centre (Thymus Chakra), where we are in touch with who we are. This connects us with the Universal Centre and allows us to be linked to another newly-activated Chakra, the Hara, it completes our wholeness. . . . The two points work closely together and as one activates, so the other commences. These provide us with a sense of balance in our lives and when the Hara drops in energy or is blocked, it affects our ability to deal with our emotional and mental problems effectively. This ultimately affects the thymus and we are then subject to infections and viruses. As the world awakens and people realize they are important and voice their feelings and expressions of self, these two chakras become stronger.[63]

—Paula Marie of The Foundation of Spiritual Healing and Guidance, U.K.

The Blooming of the Ethereal Rose

Aeons ago there were beautiful Beings who came to Earth especially to create an aroma—one which would be gentle and sweet. It would be there in Nature for all to experience and, with an innate knowing, they would be able to use it—even subconsciously—to heal their souls, to uplift, gently to nourish and fill their hearts with a wonderful feeling of well-being.

This they brought into being with the planting of a seed with great Love, and this seed grew and blossomed into the wild rose. It was white at first (which contains all colours), and as time went by, it became tinged with pink at the edges, like a gentle touch of Love. As the years and centuries went by, this gentle, wild-spreading rosebush was nurtured by gardeners; then, from cuttings, they were able to grow more roses and specialize into all sorts of beautiful hybrids. Today, there in Nature for all to see and smell, are these beautiful flowers of God.

They are filled with God Essence, which can fill us, permeating every sense and fulfilling its purpose of nurturing, healing, uplifting . . . and giving sheer delight to everyone who will take the time to pause, sense, feel . . . and for a moment relax into the aroma and be transported to heavenly realms—touched, healed and the Heart Centre filled with Divine Essence.

Be like the Rose. Let your radiance fall upon whomever draws near to your fragrance, for each of us has this, together with a Light. Let your fragrance permeate the very air you breathe in, and then release. Let it be like a very blessed mist, and those who walk within that mist shall be charmed, resuscitated, filled and healed through your beauty.

Meditation: The Bridge

With your Gossamer Golden Robes about you,
feel you are sitting by a pool of pure, clean, calm water. . . .
It's a softly warm day with a cool breeze every now and then coming off the water. . . .
You feel you are far, far away from everything . . .
suspended in space . . .
but safe, amidst the quiet and calm of Nature. . . .

As you sit quietly there, let your gaze wander to the far side of the pool. . . .
There it is hazy . . . misty . . .
and you see beautiful pale colours within the cloudlike substances. . . .
As you watch, a Bridge forms of pure, scintillating gold. . . .
See its shape . . . its firmness. . . .
You are greatly attracted to it. . . .

Let yourself go towards it . . . and then take the steps across in its shimmering Light . . .
to go for awhile into the all-enfolding Light at the far end—
the pure Golden Light of the One—and become it. . . .

Chapter Sixteen

Blue— Ray of Peace & Serenity

Let me be a reed,
And let the Breath of God
Blow through me—
Make music of me.

—Author unknown

The Light We Carry Within

Daily as we walk about the Earth, many energies are radiating towards us, enfolding, enfilling us. But also, we too are radiating Beings, and wherever we walk we impart our radiance, emanating in all directions, so that within this beautiful field of Light—whenever people walk into or through it—they are "enlightened," uplifted, invigorated, rejuvenated. In fact, because healing energies are within each of us, healing takes place if they are ready to change. That notion—a very simple idea—is so simple that we perhaps never think of it.

It is the same principle when we have a healing treatment from a source—that person is enfolded in the Light, the colours fill the Sanctuary or office, and emanate not only from their hands, but all around them. So we are immersed for a whole hour from the colours of the energy. We draw what we need for that moment, and use them to balance ourselves.

What a thought then, that we could all be channels for healing (many are, even if subconsciously)—giving AND receiving while we walk on the beautiful Mother Earth. The colour we need for today may be emanated by one special person to whom we are surprisingly (to us) drawn at that time. Remember all this as you walk the Earth—in awe, in Love and in tribute for being so guided for your daily "Light" needs.

BLUE—THE GREATEST HEALING RAY

Blue is healing. All Light is filled with Blue Rays, called the "Oxygen of Life."

The Silvery Blue Ray is for healing and teaching. It envelopes us, is quietening, calming, cooling, soothing, peaceful—it acts as a painkiller and releases tension, it helps in being detached, seeing things in a spiritual way. Blue influences the mind and the higher centres. It is the main source of energy in the Thyroid Centre, it is communication felt in the Throat Centre, and we go through the Throat Centre to the paranormal world like passing through a bottleneck.

Searching, asking "Why am I here?"—these are linked to the throat, the power of the spoken word. Blue helps in singing, and when the Throat Centre is open, then we are clairaudient. The throat is very important for spiritual development, but if the thyroid is overactive it needs Blue to slow down. We can also use sky Blue for depression.

Our Planet is Blue—it is calm and tranquil. Use the healing Blue sky as you go to work; Blue Rays are for your use and enjoyment.

Be by the sea—this lifts, exalts, inspires. The sound of the ocean calms—it is soothing to the mind and emotions.

Healers emit the Blue Ray of Peace—a happy feeling emitting loving healing energy, and those on the magnetic colour rays (Red, Orange and Yellow) enjoy the company of Blue Ray people—it calms them.

Those who are too Blue are passive, inactive and tend to put on weight—and a person who wears Blue for years is inclined to be rigid in outlook. Blue without any relief may be too depressing, self-righteous, emotionally unstable and introverting. It is suggested to wear the complementary colour of Red (or Orange) to balance.

The Blue Ray of the Soul is said to be Royal Blue—the greatest healer. It stands for loyalty and integrity; it is regal and royal—and those on this Blue Ray need to stand up for themselves.

Pale Blue, or Baby Blue, reminds us of the early morning sky. These people have a peaceful, enlightened aura. Pale Blue reminds us of the flower love-in-a-mist and is healing for the throat.

THE VOICE OF BLUE

It (Blue) is My quietness . . . My stillness . . . I fill the sky with it . . . I fill my waters with this wonderful colour. It is one of My greatest gifts to you. Fishermen and swimmers and pilots often know the Peace of going "into the wild blue yonder." . . . It calls to them in a way they cannot explain. When you feel Blue, you experience My essence . . . you partake in divinity . . . you lose your ego consciousness and experience complete bliss.

—Jules Renard, French dramatist and author: 1864–1910

Chakra #6—Throat

Location: Third cervical vertebra.
Glands: Thyroid, parathyroid.
Element: Sound.
Sense: Hearing.
Function: Speech, voice, communication, expression.
Malfunction: Sore throat, stiff neck, colds, thyroid, speech/hearing problems, communication difficulties.
Controls: Dryness of skin (thyroid), cramps (parathyroid).

The Throat Centre—Gateway to Consciousness

This centre, located in the area of the neck and shoulders, is the gateway between mind and body. This is the centre for sound, vibration, self-expression, rhythm and words. It controls creativity, transmits and receives communications. We are taking another step away from the physical—this is the first level of physical transcendence.

The Throat Centre is also affected by the teeth and whether or not they are healthy, especially if there is a problem with mercury toxicity. Remember, dentists have implanted mercury amalgam for years in thousands upon thousands of mouths, knowing it to be toxic. In the last few years, alternatives to mercury have been available and conscientious dentists have given their patients a choice—but many do not. For those readers who need to know more about in-depth studies and a protocol for replacement, a ten-page leaflet has been prepared and is available from the publisher.

The throat centre is the centre most under attack by bacteria and when the throat energy is depleted it is open to all kinds of threat—colds, tonsilitis, laryngitis, . . . an imbalance of the thyroid gland, can also lead to being overweight.[64]
—Lilla Bek and Annie Wilson

The Throat Centre relates to questions such as, "Were you allowed to speak freely as a child? Were you encouraged to trust your own perceptions? Was your intuition nurtured? Were you encouraged to feel your connection with God and Nature? Were you encouraged to sing, to express yourself freely?"

To reach and open the Sixth Chakra, the body must attain a certain level of purity. Purifying helps chakras attain greater sensitivity—telepathy comes with a well-developed Sixth Chakra. The vibrations here are faster. Meditation helps; movement of the physical body does not. We are out of the limits of time and space.

Using Blue energy creates a state of well-being and calm that often releases the need for relaxants, antibiotics and painkillers.

Using the Blue colour has a deep, penetrating effect on all four lower bodies. It causes tightening of the etheric body, which in turn strengthens the aura. It balances the emotional body by soothing the nerves, and Blue can induce sleep.

Glaucoma and cataract both respond to Blue. Gaze into Blue light or towards (not into) sunlight for thirty minutes each day.

Solarized Water
Fill a glass with water, and enfold the glass with a coloured filter of your choice (to match the relevant chakra). Put it on a windowsill in the light for twenty minutes — then drink. Solarized water can be refrigerated to keep. Blue solarized water can cool fevers.

AS THIS CHAKRA OPENS

- There will be a general feeling of relaxation, of wellbeing, of gratitude for living, of being able to express oneself easily in a kindly fashion—lovingly, tenderly, with full meaning of what is actively meant and felt. This comes about because the Heart Centre is now joined with the Throat Centre.
- The Brow Centre will be affected, thus a greater clarity in thoughts will be experienced as each life situation comes forward.
- The colour Blue in every way will be wanted—desired; at least for awhile until the centre is stabilized.
- There may at first be a need for more balance and it may be that the imbalance will be felt in the Solar Plexus Centre. These two centres can be adjusted—work with them together.
- There can be an urge to write, because communication will become so much easier—freer flowing, without fear of criticism. It will be genuine—from the Heart. The Throat is one of the greatest creative and power centres
- The response then will be to the tone of voice rather than to the actual words.

WHEN IT IS FULLY OPEN

When the Sixth Chakra is properly awakened, dis-ease will be relieved in the immune system, nervous system, throat and upper bronchials; self-suppression ends. When it remains closed, the person will be introverted, sullen and unable to articulate and express emotions.

Case History with Syntonics

Jacob Liberman quotes a case history using Syntonics (special plastic spectacle frames housing coloured gels).

> A few years ago, Robert . . . came to my office seeking possible assistance with an eye disorder that he had been told might rob him of his vision. Then 37 years of age, Robert had been diagnosed as having had glaucoma for the previous five years. Glaucoma, an eye disease characterized by increased pressure in the eye, can, if not properly regulated by medication or surgery, lead to irreparable damage to the sensitive tissues of the eye, possibly resulting in blindness. . . . Robert had an advanced case of this disease, and his left eye had lost a substantial portion of its field of vision . . .
>
> I sent him home with a special mask housing a blue-green filter and instructed him to sit outside in the sunshine every day for 30 minutes wearing the mask. I had no idea if it would help . . . I asked Robert to return in three weeks . . . To my surprise, when he returned his intraocular pressures had dropped and his visual field showed a definite improvement. . . . we jointly decided to continue with the same treatment for a much longer period of time. After four months, Robert's intraocular pressures had dropped substantially. His visual field had expanded beyond my wildest hopes . . .
>
> [Case histories like the one above] are not presented to prove that light therapy would be effective in all cases . . . but so that all of us may realize that "the sky is the limit" until proven otherwise. It is important to realize that if we don't stretch our belief systems about what conditions are treatable and what forms of treatment are effective, many of us may . . . [develop] an attitude that "nothing further can be done for us."[65]

> . . . [C]onsider what we call chemical reactions to colour, perfume and sound. This may interest people who suffer from hay fever or catarrh and other forms of chemical reaction to perfume and scent. It should be realized by those who feel the incoming perfume, scent or aroma which causes any form of irritation that, in many cases, it is not the scent or aroma . . . it is the thought picture associated with it. . . . but if we could discover the thought-force that was released when that particular odour or penetrating colour was in evidence, we could release the thought and thus free the chemical reactions on the glands, because that is exactly what hay fever and catarrh and sinus trouble is, congestion, but on a subtle level. . . .
>
> With most cases of sinus trouble we find there has been somewhere in the history of that life an emotional sacrifice, and the Soul has had to do without something that it desperately needed.[66]
>
> —Ronald Beesley

Music for the Throat Chakra

The throat area needs very spiritual music, and it is said that within the Third Eye are memories of every song you have ever sung and every tune you have ever heard. Gregorian chant or meditative music, deep and beautiful, makes one cry and releases negativity that we no longer need.

The music of Bach and Handel restores harmony—or use your hands to play the music yourself to restore Balance. This can release anger and frustration from the body.

Highly recommended are the chants recorded by Jonathan Goldman, not only for the Throat Chakra, but for all chakras: *Chakra Chants*, ISBN 0-97347-71412-7 and *The Lost Chord*, ISBN 0-97347-34032-6. Available from Etherean Music, Tel (888) 384-3732.

Case Histories with Crystal Bowls

For this chakra, Note G is needed; tone with it and remember to balance using complementary Note Middle C, while visualizing Blue. Note G most of the time gives balance of polarity and energy.

When the Note G Bowl was played, I immediately felt vibration in my hands. It sounded screechy and discordant. The tone was the same as my tinnitus, but all discomfort is now gone. —M.B.

C, D, E, F, F#, G, A♭, B, high C were played. The sounds affected me immediately, going directly into my energy field and physical body. They were felt in my neck/shoulders/head—very calming, peaceful and balancing. The pain in my neck and shoulders has subsided. —T.P.

Client has backache. Scan showed canal blocked; needs surgery to clear. X-rays showed discs were damaged—the pain goes everywhere. Used Magenta lamp on neck, Orange lamp on lower back. We toned together and played Bowls, especially G near neck. Felt great and out of pain. —A.B.

Mary McCandless in Toronto sang "Silent Night," then played the Bowl in the same key as that of the song (Note G). This had a profound effect on a group of parents and friends of lesbians and gays. There was an overwhelming response—some were moved to tears. The process of communication and opening up among strangers seemed to have been triggered. The room became very peaceful.

Speak Your Truth, Speak Your Peace

Many of us who walk the Earth now have come to lead others to the true pathway of Light, where we can commune easily and truthfully, and be peaceful, one with the other.

However, it is true that earlier we held our Peace—held our tongue, for that matter—and only the very daring ones would speak out loud the Truth. But as time marches on so speedily, there is now more openness to speaking AND listening. The wiser ones can speak openly, fearlessly and more loudly for others to hear and heed what they need to know—and to then realize that these Truths and feelings are already within their knowingness at the Heart Centre. They can then connect, feel relief and Joy all at the same time, because they can act upon these true thoughts and be a more open channel to help and encourage others on this great Pathway to Truth and Peace.

We encourage you then, when asked, to openly state your Truth, but in very simple terms, with fewest words. Let them drop like pearls, only one at a time, to nourish, encourage and above all be assimilated, then lived!

Now is the time to speak your truth; now is the time to send this Peace before you—for indeed, it IS felt. Now it has to be expressed with Love.

Next time then, when questions are asked of you, open your heart. Allow the Joy to be released, to no matter whom, and ask to be helped to give the right answer, but briefly—then leave it. In time they will return for more. Let your Light shine. There is a great search for beacons along the Pathway—be one of them, even in small ways, which are the best, and lay a very firm foundation along life's Path. Be that lovely Light and feel the Peace.

Meditation: The Bluebird of Happiness

You are in your favourite garden, lying on a sunny spot on the green, green grass. . . .
It is a drowsy kind of time, with the hum of bees and crickets . . .
and you are contemplating the blue, blue sky . . .
clear . . . cloudless . . . beautiful. . . .

Into your lazy gaze comes the most delicate Bluebird . . . small . . . exquisitely formed . . .
with deep blue headdress . . . and shaded blue over his body. . . .
He sings with a full-throated song his Joy and delight
in the wonderful day in Nature. . . .
The sky is his world. He is in tune with it . . .
his colours match . . . and he shows you how to be free . . .
to soar upwards and away to more glorious climes for refreshment . . .
for serenity . . . for Peace. . . .

He is the Bluebird of Happiness. . . . Be him in your Heart Centre. . . .
Feel the freedom . . . the stillness . . . the real YOU. . . .
All things are possible. . . . Let go . . .
and let his song carry you on Wings into the Blue. . . .

Chapter Seventeen

Indigo— Ray of the Aquarian Age

If therefore thine eye be single, thy
whole body shall be filled with Light.
—Matthew 6:22

The Magic of the Eyes

When we feel lonely, sad, neglected, uncared for, there comes something like a film over the eyes—as if we are not seeing properly, and everything is blurred, hazy. We feel sick and have this great longing within.

Longing for what? It is the great Sun—the beautiful Spiritual Sunlight which permeates all things, if we will allow. We need not go into the mechanics of Light entering through the eyes, reaching through to the optic vision and to the pineal gland. Nevertheless, this IS the Pathway to Light, life and wellbeing—and we do need to know this, accept the idea, get out into the air, the Light, the Sun, and breathe it in, soak it in. Such a feeling of upliftment, wellbeing, Joy ensues. We then feel better, feel good and know inwardly that we are cared for with Love and understanding.

Light is really all that matters. It fills us, it has all colours—reaches into every single cell. The colour we need momentarily (perhaps Blue or Indigo for the eyes) has health and nourishment. It reaches into our bodies, touches every part of us, every single hair on our head, every single particle and cell—nurturing, healing.

All we need to do is ACCEPT the Light into us, give deep thanks, breathe it in and release with the breath all our doubts, fears, guilts, worries and leave them behind. Here is Joy, wonderment and caring beyond our belief!

Allow it then—allow the Light to permeate your whole Being. Be the Light and be thankful.

Indigo—The Soul Colour

Indigo is a powerful ray; it is the Midnight Blue of the heavens—an electric, cooling colour with a very healing quality. Indigo is the Soul colour—it represents the search for the inner you; it is related to spiritual understanding.

Indigo will recharge all cell tissues. It is purifying and anti-inflammatory; it cleanses and burns away toxic residues. It links the Third Eye with the pituitary gland, and links to the skeleton. Indigo can help close the door on the past, and it helps to clear the mind.

Indigo Ray people emit a lovely, peaceful, healing energy. They come in consciously aware from the time of birth. These children (souls) find it difficult when young—boring—for they have come back to be a teacher of parents (our children are our second education!). See the book *The Indigo Children*, by Lee Carroll and Jan Tober. (See Chapter 4, page 35 for excerpts from this book.)

Cataract and glaucoma have responded to Indigo: shining Indigo light on the head and eyes over a period of a few weeks, has restored partial eyesight to a "dead" eye. Also, bathing and breathing in the Indigo light helped a woman to clear partially the cataract—her eye became more vital.

Chakra #7—Third Eye

Location: Centre of the head, at eye level, behind the forehead.
Gland: Pineal—connected to the eyes.
Function: Intuiting, "seeing."
Malfunction: Blindness, eyestrain, blurred vision, headaches, nightmares.

Works with: Spleen Chakra (#2—Orange).

This centre:

- stimulates the pineal gland, for creative visualization—visions;
- is linked to spiritual and soul bodies;
- is the centre of intuitive seeing.

The Third Eye is a tool of the Seventh Chakra. Sounding AOUM, with special emphasis on the "MMM" sound, vibrates the bones of the face and stimulates this centre.

Music for the Third Eye Chakra

- *Piano Reflections*, Kelly Yost (Channel Productions, P.O. Box 454, Twin Falls, ID 83303. Tel (208) 734-8668
- *Garden of Serenity*, David & Steve Gordon, Sequoia Records, Box 280, Topanga, CA 90290.

People on the Indigo Ray have a natural inclination towards the healing services (e.g. many doctors and nurses); they have a need to help people back to health. It is a quality of mirroring back what we are. Indigo people have a stillness inside them, and the greater the stillness, the more ethereal the colour.

People on the Indigo Ray like to wear Blue. They appreciate sacred music; they work in alternative medicine and are drawn to Yoga practice.

A depletion in the Brow Centre means we meet stress on a more psychic level. This area registers the effect of the "unseen" things—like changes in the weather, phases of the moon, earth's vibratory rates—on the body. All will pull at one's energies.

Sometimes the Brow Centre opens before others—but it will not then function as well until the Heart Centre too is flowing and gently pulsating. The Brow Centre also works with the ones above and below it—Crown and Throat.

So it is best to do exercises each day to encourage these centres—yogic breathing is excellent.

As the consciousness becomes increasingly focused on awakening of the heart-mind, a feeling of brotherhood for all life begins to develop and the higher chakras in the throat and head begin to quicken. The safe and harmonious way for these to develop is through increase of human and Divine Love in the Heart Centre. No meditation or spiritual exercises are really effective until the Heart Centre opens in Brotherly Love and service to all life.

LIGHT through the optic nerve feeds the autonomic nervous system:

- we are totally dependent on this energy;
- glasses, sunglasses, contact lenses screen out light;
- some are exposed all day to artificial light (no colour frequency);
- chakras are ports of entry and exit for energies;
- the Sun is the symbol of life and Light;
- Light is the energy of Joy;
- our work is to get into the LIGHT.

When thoughts of a greater frequency come in, they are taken through the awakened portion of your brain. The pineal gland in the back of your head receives the greater frequency, and it begins to swell, which causes your head to ache; or you may feel a little dizzy or light-headed. That frequency is then turned into a high-powered electrical current and shot through your central nervous system to every cell of your body. From that, you will feel a rush, or sensation of tingling, of being lifted, because greater energy than you've felt before is now rushing throughout your body. The frequency ignites every cell, causing it to increase its vibrational frequency. The more you receive unlimited thoughts, the greater the body vibrates, and you begin to take on a glow—because you are beginning to reverse the body from density back into Light.[67]

—Barbara Ann Brennan

AS THIS CHAKRA OPENS

One might experience:

- being out-of-body slightly; also some dizziness or some pressure across the forehead, or in the head;
- some cloudiness of eyesight;
- images before the sleep state begins;
- stars before the eyes;
- light seen often to the right of the eyes when closed;
- seeing the auric colours around the body, not always complete but here and there on some very luminous people;
- a feeling of "knowing" without reason and being aware in a new way.

Above all, fill yourself with Light to protect yourself from any disharmonies. Usually it is a gradual opening, but this can depend on past-life experiences of use of the Third Eye.

WHEN IT IS FULLY OPEN

- above experiences will clear;
- there is an expanded awareness;
- clairaudience is also experienced sometimes—like the "knock on the door" (see sidebar).

One needs to keep oneself balanced in all ways, because when the expression becomes more fluid, the energy from the Throat then spreads to the Brow and it begins to become illumined. When, in turn, the Crown Chakra opens, we breathe in Light energies which interact as they flow down the spinal cord. This strengthens nerve responses in the body, the muscles and the working of the endocrine system. We are then more able to command our lives.

As one is opening spiritually, auras and chakras are seen and details of the astral plane, as well as events in other times. We become able to see into other places and everything invisible, such as Nature spirits. We see ugliness beautifully—as part of an evolutionary process (e.g. Mother Teresa assisting the dying in the slums of Calcutta, seeing the Christ force in every person). When one achieves such a state, one needs just to be, to accept, and especially to protect oneself at the beginning of each day.

The Bell or Knock

When our daughter Fiona was taken to the hospital at age two or three (for what, we didn't know exactly), tests were taken during the night and the staff said we would be informed as soon as results were known. In the middle of the night, I awoke suddenly hearing footsteps up the gravel drive and a "knock" on our front door—then footsteps dying away, retreating down the drive. I "knew" not to go down and answer the door. I "knew" no one was there, and I also intuited that the doctors had found the cause for Fiona's sickness. Next morning at the hospital, they said it was a simple bladder infection. Once more—thank you, God! (My healing circle had sent much Light to the situation.)

Gravity/Magnetic Balancing

As one begins to spiritually open, dizziness may occur. This suggests one is slightly out-of-body. Our lymphatic glands may be acting up; our body may need inner cleansing; we may be suffering from digestive pressure. These all may be occurring because:

- blood-sugar level is low;
- blood pressure is low;
- we are psychically overextended (from too much healing work and healing groups).

Breathing techniques and physical exercise will help. There is also something called Gravity/Magnetic Balancing, developed by Ronald Beesley to balance the bodily energies—many have found it to be effective.

Colour has a powerful effect on you. The colour that you wear closest to your hypothalamus (a gland located in your head) has the greatest influence. A wonderful inexpensive form of colour therapy is to have several different coloured pillowcases from which to choose so that each night you can find the one that most enhances you.[68]

—Jeanne Elizabeth Blum

TO GIVE A GRAVITY/MAGNETIC BALANCING

Stand behind the person being treated, who is seated comfortably with feet flat, relaxed. Comb down their energy field with the tips of your fingers barely touching their body, from top of head to base of spine twice. This is very relaxing. . . . Then follow the instructions below:

- With your hands held at either side of their body, feel for the gravity field at elbow level or hip level. This may feel "sticky" in your hands—or hot or tingling.
- Slowly bring up this energy field with both hands (may feel heavy) until you are above the head.
- With left hand, palm down, hold field there.
- With right hand, reach up above and feel for the magnetic field, which has gone upwards away from the body (usually held by the gravity field); it may feel like a "shelf."
- Bring the field down slowly to the left hand and "tie" together (hand over hand)—once is sufficient.
- Comb down the spine once more.
- This can be followed by a chakra balancing (see *THTCB*, page 90).

For further help with grounding and relaxing, nerve salts—Mag. Phos. and Kali. Phos., available in health food stores—can be useful. Dissolve twelve of each in a tumbler of water and sip during the day.

Case Histories with Crystal Bowls

For the Indigo Chakra, the Third Eye, we need to sound the Crystal Bowl A (above Middle C) while toning the same note. Alternate with complementary Note D for true balance, visualizing Orange.

All Bowls were played—sound was felt immediately in the head and heart. The effect was an expansion of consciousness, very soothing and relaxing, and my back feels better. —M.D.

I use Reiki in my treatments, and use the Bowl a great deal too. More sensitive people say the sound of the Crystal Bowl helps to release old energies stuck in their auric field. We also use the Bowl on the water in the bath and play it—a wonderful experience. We used it likewise for someone with intestinal problems and it helped a lot. Children also love it at bath time; it is calming for them. —Godelieve, Belgium.

A♭ BOWL—INDIGO:

This person gets really tired; her cheekbones hurt, she had a slight rash on the face, pain in the neck and inflammation of the nerves after a virus. Her family was planning to return to New Zealand the following week. Indigo oil was used on the face; we toned together and played the Bowls. She wrote two months later from N.Z. that ALL pain left after the treatment, and has not returned. —L.S.

Debbie Cottle, massage therapist specializing in Bioenergetics in Salt Lake City, Utah, reports: "The A# Bowl was being demonstrated, when after about two minutes of toning, a woman present broke out in a bright red rash. It was yeast exuding through her skin. The same thing happened a year ago, and the A# Bowl has been used on several occasions since on people that have yeast problems; it has been helpful to them."

Cottle further reports: "The A# Bowl has been a great help on several occasions with people dealing with severe negativity . . . because of aligning the Third Eye chakra . . . which is vital . . . to allow a positive spiritual alignment."

—Debbie Cottle,
Email: <Lighttouch2000@aol.com>

Meditation: Still, Still Waters

Still, still waters . . . calm and steady . . . cool and deep . . .
very clear, reflecting the Sun and sky and trees. . . .
Joyous birdsong clear on the mountain air,
almost like a message: "Be still, listen, sing with me. . . .
How beautiful is life. . . . Take time to listen to the song of the Spirit . . .
to be uplifted . . . to join with all Nature in praise of Being. . . ."

Enjoy each season. . . . See the seasons in you . . . the growing in the dark of winter . . .
the nurturing of the Spirit towards the growing Light of beautiful, glorious Spring . . .
when all buds burst forth in the Sunlight . . .
the Spiritual Sun which gives such warmth and encouragement
to all growth of the soul.

Reflect on the summer Sun shining on the cool, deep water . . .
on how there are great depths within you . . . and the warmth of Love
from the Spirit (Sun) can bring these beautiful depths to the surface . . .
so that in the Autumn of Life we can share this warmth,
we can shed our "leaves" of beautiful colours . . .
to give Joy to others, protection and the wisdom that each leaf carries within . . .
food for the Earth, but also food for thought.

And as we gently reflect on all these thoughts . . . by the still, calm lake . . .
we are enfolded by a wonderful, gentle Peace . . .
so that nothing else matters. . . .

We are at Peace with all life. . . .

Chapter Eighteen

Violet—Ray of Spirit

To thine own self be true,
And it must follow, as the night the day,
Thou canst not then be false to any man.

—William Shakespeare's *Hamlet*

'To Thine Own Self Be True'

Have you ever pondered on this quotation? Thought about the Truth within yourself—how to apply it, and what does it mean?

Throughout the ages there have been numerous Sages, at whose feet sat many students. These Sages, each in their own way and language, passed on choice "sweetmeats" so that a student had something on which to ponder, to question, to try to evaluate and use for himself.

These sweetmeats were really supposed to take the student within, to the temple of his heart, to use in a gentle way, to get to the very heart of the matter, in fact—in other words, to find a true understanding of oneself. Then, in this way, could one carve out a pathway for oneself, knowing exactly what is—and what is not—beneficial for one's own evolution; what suits one person will not be right for the next.

And so in Truth, it is a delving within: How do I feel about THIS? How do I react to this idea? Is it good for me? If I use it, will it help me to evolve spiritually and to then help others?

If I don't care for the idea, shall I then discard it immediately, cleanse myself, or dabble with it until it becomes diffused, and maybe some of it "sticks" with me when I don't really have a need for it? Can I train myself to watch and be aware of my thoughts, my feelings—especially those two—and to allow myself to be very aware of what I "take on," latch on to? Can I be really strong and reject, but kindly, that which I really don't need, to put it out of mind, and continue with gentle thoughts, so that I am not cluttered, my solar plexus is not in a "knot," and in this way I have immediate understanding of myself?

I can actually conduct myself with understanding and Love, and allow myself to be a clearer channel for uplifting thoughts, kindness and compassion—those are the healing, cleansing thoughts (not to be invaded by thoughtlessness)—from myself, but from others too.

In this way can we keep ourselves healthy and clear, walking with head held high, Heart Centre open, not "holier than thou," but truly understanding and loving oneself, being true to that inner core of our Being—therein lies Joy!

Chakra #8—Crown

If I were to begin life again,
I should want it as it was.
I would only open my eyes
a little more.
—Jules Renard,
French dramatist and
author: 1864–1910.

More than any other colour perhaps, it is Violet that acts upon the subtle bodies. Violet light can help dissipate negative astral entities. It is a colour that can dissipate, transmute and change blocked conditions, and open channels to correct energy flow and balance.

Violet is the highest-vibrating colour of the visible-light spectrum—it can quicken material substance and change it to more etheric and subtle expression.

Violet is the great transforming power of the New Age.

A Life for A Life

Every day we live and die, all in the same day. What does this mean? We have life, we breathe—but we also breathe out, and that is the death part we mean. We breathe in the breath of life, that we may nourish ourselves on all levels, physically, mentally, emotionally and spiritually—be filled with the Christ Light that all may be so illumined, and therefore, those whom we touch are also illumined.

And then having nourished ourselves, the negative parts of us we must allow to shrivel and die, to be released on the outbreath. If we think about it, it seems to be very simple, and so it is. We ourselves make it difficult by our thoughts; we forget to breathe consciously every now and then, to release. If we did, there would be no problems, no dis-ease, no discomforts; we simply would be filled with the Light of the Sun, and all our cells would sparkle—we would feel so WELL.

The simple trick in overcoming this is to change our thoughts—when we feel on one track constantly, simply go out into Nature for awhile and breathe deeply, breathe in the JOY, fill ourselves with it, and as we exhale, allow the release of everything unwanted, until there is only room for the good breath, the Light; then all else is squeezed out.

Use the Golden Sunlight in your mind's eye then. From now on be filled with it every time you take a breath. Give thanks for it and for this life; then, when there are no more negativities to exhale, we can only breathe out JOY wherever we walk.

Balancing the Astral/Emotional Body with Crystal Bowls

The positions shown for the Crystal Bowls in Figure 6 are for toning the fifth intervals, beginning with Note F at the base, then C to G, D to A, E to B, followed by F#.

Playing them in this order, with the patient lying comfortably in the centre of the circle, balances the Astral body, therefore healing the Emotional body, and the resonance returning from the sounding is equal among all Bowls.

When the sound comes back right away, like an echo, then the aura will be harmonized. Afterwards, it is suggested to "fix" the effects of the sound and to balance the general energy using Orange and Blue lights, or coloured glasses.

Begin by sounding Note F at the feet, then progress to Red Note C, to fifth-interval Blue Note G, thence to Orange Note D (Sacral Chakra) to Indigo Note A (Third Eye), to Yellow Note E for the Solar Plexus, to the top for Violet Note B Crown Chakra, finishing with Turquoise Note F# Thymus Chakra. There have been some remarkable results.

It is interesting to note that Complementary Notes C and G are Red and Blue, which together make Purple, and this creates a very high harmonic—which is most healing. Sounding fifth intervals helps to change the vibrations in a room!

The Crystal Bowl to sound for Violet or Purple—the Crown Chakra—is B. Tone with it and also its complementary Note E, visualizing Violet, to achieve balance in every way.

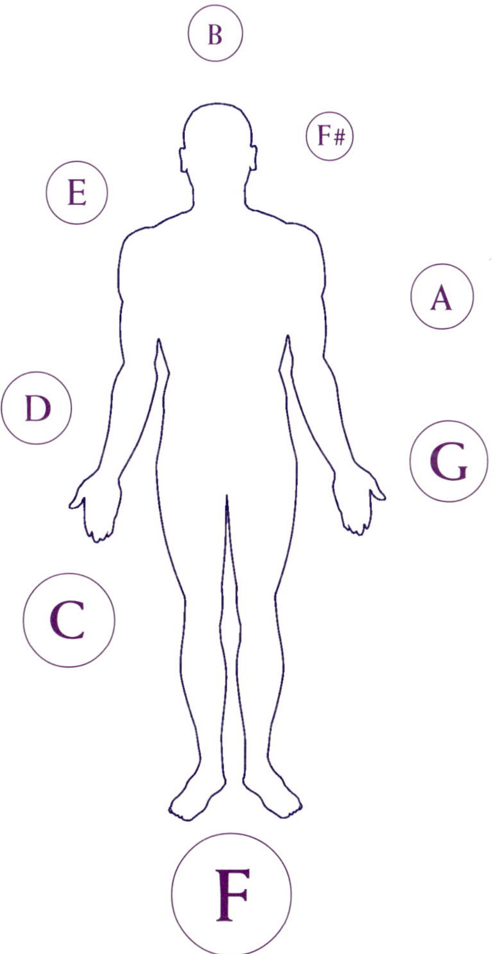

Figure 6.[69]
Positions for Healing Bowls

The Process of Death (or Rebirth)

Barbara Ann Brennan, in her book *Hands of Light*, writes:

My guide has been lecturing on the death process, and I would like to quote him here. First he says that death is not what we have understood, but a transition from one state of consciousness to another. Heyoan says that we have already died, in forgetting who we are. Those parts of us that have been forgotten are walled off from reality, and we have come into incarnation to retrieve them. So although we fear death, he says we have already died, and in the incarnation process of reintegrating with our greater being, we actually find more life. He says that the only thing that dies is death.

During our life, we wall off experiences that we wish to forget. We do this so effectively that we do not remember many of them. We begin this walling-off process early in childhood and continue it throughout our life. These walled-off pieces of our consciousness can be seen in the auric field in terms of blocks . . . Heyoan says that the real death has already occurred in the form of that internal wall.

"As you know, the only thing that separates you from anything is yourself. And the most important thing is that death has already occurred in those portions of yourself that are walled off. That would be perhaps from our vantage point the most clear definition of what the human being considers death. It is being walled-off and separate from. . . . It is forgetting who you are, that is what death is. You have already died. You have in fact incarnated to bring to life those pieces of you that are already in what you call death, if we should ever use that word. Those parts have already died.

"The process of death, that which we would call transition to greater awareness, can be seen as a process in the energetic field. We will describe this now to help you understand the process of death from the auric point of view. There is a washing of the field, there is a clearing, an opening of all the chakras. When you die, you are going to another dimension. There is dissolution in the three lower chakras. There is dissolution of, and note we say dissolution, of the three lower bodies. Those of you who have watched individuals die have seen the opalescent quality of the hands, of the face, of the skin. It is an opalescent mother of pearl as the individual is dying, and the beautiful opalescent clouds are wafting off. Those clouds are the lower energy bodies which serve to hold the physical body together. They are disintegrating. They waft off, and the chakras there are opened and there are cords of energy coming out. The upper chakras are great open holes into other dimensions. So this is the beginning stages of death where the energy field begins to separate. The lower parts of the energy field separate from the upper parts. And then during the three hours or so around the hour of death, there is a washing of the body, a baptism, a spiritual baptism of the body where the energy is flushed through like a fountain right up the main vertical power current. A fountain of golden light flushes through and all of the blocks are cleansed. And the aura becomes white gold. How will this be experienced by the individual who is dying in terms of memory? You have already heard it. A person sees his or her entire life wash by them. Well, that's it. There is a concomitant energy field phenomena of the washing of the aura. All blocks are let go. All forgotten experiences of that lifetime are unblocked. . . . Thus, all of the history of that lifetime flows through the consciousness and when the person leaves, the consciousness leaves. It is the dissolution of many of the walls that were created for the process of transformation in this particular lifetime. It is a tremendous integration.

"With the dissolution of the walls of forgetting within you, you remember who you truly are. You become integrated with your greater self and feel the lightness and the vastness of it. Thus death, contrary to popular opinion, is a very wonderful experience. Many of you have read descriptions of those who have been pronounced clinically dead and have come back to life. They all speak of a tunnel with a brilliant light at the end of it. They speak of meeting a wondrous being at the end of that tunnel. Most have a life review and a discussion of that life with that being. Most reveal that they themselves decided to return to the physical world to complete their learning, even though it was so beautiful where they went. Most of them no longer fear death, but look forward to it as a great release into serenity.

"So it is your wall that separates you from this truth: What you call death is actually transition into light. The death that you imagine you will experience can be found within your wall. Every time you separate yourself in any way, you die a small death. Every time you block your wonderful life force from flowing, you create a small death. Thus as you remember those separated parts of your being and reintegrate them back into yourself, you have already died. You come back to life. As you expand your awareness, the wall between the world, the wall between spiritual reality and physical reality dissolves. Thus death dissolves, it is nothing but the releasing of the wall of illusion when you are ready to move on. And who you are is redefined as the greater reality. . . .

"Your physical body dies, but you move into another plane of reality. You maintain that essence of self beyond body, beyond incarnation. And when you leave your body, you may feel yourself to be a point of gold light, but you will still feel yourself."[70]

Music for Rebirth

Finally, when the time comes for a soul to leave the beautiful Mother Earth and return to their Home in the Light, we can help them onwards by having uplifting music at their service. Music will help them surrender their physical bodies more easily and not worry too much about leaving, especially loved Ones, but to be able to release the body of Light from the physical body and return to their place in the Dimension of Light.

Here are some suggestions for suitable music. It is always a matter of individual choice, but whatever music is chosen, try to select melodies that inspire, lift and lighten the atmosphere. The loved one's passing should be a time of great Joy—play music that celebrates their transition as a true graduation and victory.

- Schubert—Ave Maria
- Richard Strauss—Transfiguration (closing section)
- Wagner—Pilgrim's Chorus (from *Tannhäuser*)
- Humperdinck—Children's Prayer (from *Hansel and Gretel*)
- C. Bohm—Calm as the Night (organ)
- Fauré—In Paradise (from *Requiem*)
- Herbert—Ah, Sweet Mystery of Life
- Elgar—Pomp and Circumstance
- Mendelssohn—Selections from Elijah
- Brahms—How Lovely is Thy Dwelling Place

These are uplifting for everyone. We can use colours that will help, too—like Violet, Rose, Gold and White. Fill the place with lovely flowers of those colours, and know how helpful this can be to all those who gather to acknowledge that person's life. Breathe in the Light together and send forth the Light to that Soul on their journey Home.

An excellent book for those seeking help overcoming grieving is There is No Death. It is a beautifully written volume, by a woman who lost two sons when they were quite young.

There is No Death
By Betty Bethards
The Inner Light Foundation
P.O. Box 750265
Petaluma, CA 94975
Tel (707) 765-2200
Sept. 1990 (Fifth Printing)

Here is another book suggestion:

Journey of Souls
By Michael Newton, Ph.D.
Llewellyn Publications
St. Paul, MN 55164-0383
1994 (Eighth Printing 1999)

Resting Assured that Life Goes On

This is a very big question with many people, but what we need to understand is that life here on Earth is very minimal compared to the Great Beyond, or the next plane of Light. We are here seeing through a glass darkly, and whatever befalls us we have ourselves ordained. What if we fail? What if we do things very differently? Does it matter? Isn't it all experience, to be accumulated in a book somewhere far off? Is this what it's all about? How do we know for sure?

How about when people leave the Earth, what do we say—how happy we are for that person? Is this hypocritical? But we are, in fact, so happy that they are out of pain, out of misery—and yes, of course, we'll all meet again if we are meant to do so. And what of those left behind? Are they grieving for that person who has gone ahead, or simply for themselves—for unsaid questions, unsaid apologies, for not making the Way clear and uncluttered, but only giving Love?

Case Histories with Crystal Bowls

I was lying comfortably on the floor with the Bowls around me while they were "played." The Root Chakra vibrated and I felt the sounds entering all the cells in my body—it was so relaxing. My throat tickled—I wanted to laugh for how relaxed I felt all over. I must have gone through a lot of healing. As I sit here now, I realize the pain in my left leg, hip and shoulder has gone. —I.Y.

The sounds permeated my whole Being, and I seemed to float away for a long time, feeling so relaxed, calm and peaceful. All anxieties left me, and when I came to, realized that really I have no problems—life is good. —D.F.

I joined a partner for the Astral Balancing circle. At first I was very aware of the Bowls being played, but when the top Bowl, Note B, was reached, I seemed to be in a wonderful cloud of Violet and the words came to me several times: "I will never leave you. . . ." I felt absolutely marvellous. —D.R.

Time Marches On

We tend to think of time as here and now and slipping through our fingers and as we grow older, that time seems to go more quickly, and there is hardly time enough to fit in all the things we feel we would wish to accomplish.

We need another view of all this, to, as it were, take ourselves off the beautiful Planet into the atmosphere of the next plane surrounding the Earth (for planes are never far away, but intermingled). From there we can take a bird's-eye view—see very clearly that time IS, it goes on, it can be all things at the moment, past, present and future. If we can see backwards and forwards at the same time, then the moment takes on new meaning. We can accomplish all we need in the moment, knowing that there is more time always than we need, for life goes on into and from eternity—we can take all the time we wish.

However, it does not mean that we can WASTE time. It means simply that the moment is all that matters and we must make the most of that precious moment, being true to our very innermost. We must try to do all we can to BE in that moment (not wishing to be somewhere else, doing something else, be with someone else); instead to enJOY and allow that Joy to infiltrate not only our whole Being, but the atmosphere too—the room, the place, the environment—for your Joy can "colour" others and allow them to bring forth their enJOYment too.

In this way, time as such is used, never wasted, and life can proceed in measured tread, with Joy marking every single footstep along life's Pathway.

Music for the Crown Chakra

- Greensleeves. This melody affects the soul, the whole body and touches on immune deficiencies.
- Baroque music by Vivaldi.
- *Chants by the Benedictine Monks of Santo Domingo Rosa Mystica*, by Therese Schroeder-Sheker, Angel Records.
- *Vox de Nube*, by Noirin ni Riain and the Irish Monks of Glenstal Abbey, Sounds True, 735 Walnut St., Boulder, CO 80302, 1-800-333-9185.
- *A Feather on the Breath of God: Sequences and Hymns by Abbess Hildegard of Bingen*, by Gothic Voices with Emma Kirkby, Harmonia Mundi, 2037 Grandville Ave., Los Angeles, CA, Tel (310) 559-0802.

Meditation: The Mists of Time

Create a Vision of your whole body as a great cloud of glorious Light. . . .
Feel it pulsating as you gently breathe in and out. . . .
Feel, as you inbreathe, a glorious web of intricate weavings of colour . . .
and as you let go of the breath, the cloud emanates a beautiful quality
of gentle energy . . . growing larger with each exhalation. . . .

As you continue to do this, you feel yourself beginning to move,
slowly at first, then gradually upwards . . .
and as you rise into the Cosmic Spheres, you feel such an upliftment . . .
a gentle soaring into space . . . a pulling towards other Galaxies . . .
outer space, and yet you're still part of the one whole Atmosphere. . . .

You feel so much a part, but knowing how infinitesimal your own space is,
you feel greatly the immense Joy there is in being ONE with the WHOLE . . .
an indescribable Peace . . . Belonging . . .
as your small cloud joins to the Mists of Time. . . .

And in the Atmosphere, colours and music intermingle,
touching every cell in a Heavenly Healing. . . . We can just BE . . .
but at the same time allow this great flow of Harmony through us . . .
through our cloud . . . the mists . . .
and in the experiencing become Perfect Peace. . . .

Chapter Nineteen

Magenta— A Master Ray

Any kind of disease should be considered as an alarm signal for disharmonies in spirit and soul. . . . We have to turn away from our suffering and sickness, because they are the hindrances on our Path towards the Inner. We have to bid farewell to those frustrations and fears which prevent us from progressing and from coming into contact with our Higher Selves.

—Peter Mandel, outstanding German scientist and healer, creator of Colourpuncture, said to be part of the medicine of the future.

Connecting Daily with our Higher Self

During ordinary, everyday chores—when we are cleaning the kitchen sink, making beds, dusting, washing the car, sweeping the driveway of leaves, gardening—we need not think too hard, and therefore, our thoughts and minds being free for the moment, our Higher Self can "slip in" as it were, feed in some helpful thoughts to use on our daily pathway of discovery. Yes, this can happen in meditation too—it is a linking, which can be conscious—but it often happens when we are not concentrating, and then we are surprised, happy and very thankful—"Oh yes, of course, why didn't I think of that before—thank you, dear Higher Self"—and we carry on with our current task.

Our Higher Self is there, constantly alert for us, hoping to find a space in our train of thought, to connect helpfully. Let us then endeavour to cultivate this connection—a wonderful communication there for each of us if only we will listen, make it a daily practice to be alert and ready, jot it down before we forget—and be grateful always to our beautiful Higher Self.

The Magenta Ray

Magenta is more important now because we are in the Aquarian Age. It is a higher spiritual force working out of the Red Ray that has the quality to dissolve, let go and overcome, transform dross and anger into Joy. Magenta gives upliftment and helps us to move into the right place. It is the colour of spiritual energy.

Always remember: Symptoms in the body are the only way that the Soul can get our attention!

Magenta helps the intellect—we need it in our aura if organizing or administering. Breathe it in, wear it, think it! Magenta has a magnetic quality and, used on the Gateway at the base of the spine, it animates and enlivens.

Magenta is the colour that works most efficiently on the Etheric Body. It also balances the Emotional Body, and by working on these two bodies, it builds and strengthens the Aura. It raises or lowers the blood pressure—whatever is needed.

It is the Magenta Ray which is now coming in for the Age of Organization. It is said to spiritualize the Red.

This is your mission, my brethren, to love . . . to love without criticism and without compromise; and to love means to give and to serve and to understand.
—White Eagle Calendar, December 1999.

Oh! To the God of Love we have prayed to give us Love:
Love in our thinking
Love in our speaking
Love in our doing
Love in the hidden places of our Souls
Love for our neighbours near and far
Love for our friends old and new
Love for those with whom we find it hard to love or like
Love of those with whom we work
Love of those with whom we play and feel at ease
Love in joy, love in sorrow
Love in life, love in death
Love as life eternal and eternally.
—Helene Harris, Ph.D.

Case Histories with Crystal Bowls

We must remember that the Crystal Bowls are simply tools to take us to a higher level of awareness—the sound and vibrations help to clear the energy fields. Alternate with toning this centre's complementary note—F (for the Heart Chakra!)—while visualizing Magenta.

One can place the Bowl between oneself and a friend during a conversation—many times it has been an honour to experience the Bowl sounding when speaking of spiritual truths. In New York, Dr. Mitchell Gaynor, a cancer specialist, plays a Bowl while talking to a patient and says there have been some remarkable results. He places it on the desk between them.[71]

Note High C (Magenta) was played. I heard all Bowls, but especially the High C. I felt it in the head area and all over the body—it was so peacemaking. —J.S.

All Bowls were played, but especially F, F# and High C. This person was overstressed, had hypothyroidism and high blood pressure and was on diuretic tablets for two to three years. We used Magenta oil with the Magenta lamp on the heart area, toned together and played the Bowls. She felt warm tingling in her arms and hands—became relaxed. I suggested taking the Magenta oil home to use for high stress. —K.V.

She has pain in feet, feels stuck in life. We used Magenta lamp on hands, lower legs and feet. We played the E bowl below her feet, toned together and then played all Bowls including high C. No pain in feet after, and only some in left hand. Will use Magenta filter at home for one week, morning and afternoon. —C.S.

Hurt left kneecap in winter (it's now March) and is low in energy. Used Magenta lamp and oil on knee, Red lamp on feet and soles for energy, plus Red oil. Played most Bowls and toned together—he did overtoning. Afterwards felt very light and all pain gone. —D.M.

Music for the Astral Chakra

- Rodgers and Hammerstein—Climb Every Mountain (from *Sound of Music*)
- Grieg—Piano Concerto in A Minor (2nd Movement)

Our Magnificent Soul Body

What is the Soul? What part of us does it occupy? How can we become part of it—or are we already so? How do people recognize the Soul part of us? What can we do to develop it, to become conscious of our Soul body?

These are manifold questions that people have all the time, and we shall try to explain for you. Take what you need for the moment and allow yourself to meld with it. Realize that we are a many-splendoured Being—we have many layers, the most dense being the physical body which we can feel—pinch yourself, and FEEL!

Then there is the Etheric Body, which exactly duplicates the dense body and is a lighter vibration. It pulsates in and out of the physical and houses the energy centres (the chakras); without this energetic body we would not be here, for it energizes us—fills us with prana or life force.

Next we have an Emotional Body—a slightly higher, faster vibration than the Etheric. It is Pink, carries all our emotions, and links in to many people, places and planes. On the next level is the Mental Body, linked to the many thoughts and ideas of the mental plane, coloured Golden Yellow—a faster, higher vibration than the one before. It is a stepping-stone for the Astral Body, lovely Green, which touches the Astral plane and links with entities there. We need to be reminded always to aim for the highest in our searchings, to the Higher Astral realms, where there are beautiful Beings.

Each of these levels is a faster vibration of energy than the one before, and the Astral is the stepping stone to the Soul Body—again a faster vibration, very beautiful Blue. This is really the real you—your Higher Self, there at all times, sending lovely messages to help in every way, waiting very patiently for you to link in, hoping you will make this link many times during the day.

In the sleep state, however, subconsciously we join with our Soul Body, or Higher Self, to travel to the Spiritual plane to refresh ourselves, to renew, recharge, enfill ourselves with the Light, in readiness for the next day and waking hours.

Ponder on these things, have Joy in the thought of your magnificent Higher Self being there at all times ready to ensoul you, refresh you, just for the asking—and bless yourself!

How Do People Recognize the Soul Part of You?

The very first actual contact we have with people is through the eyes (we know they are the Windows of the Soul). Whenever we see into the eyes, we immediately feel the quality of the Soul—soft and Soul-caring, compassionate. We see the integrity, we feel the steeliness, the hardness, the glitter in the eye and we have an immediate feeling of alrightness, or not alrightness.

This is the way we feel and recognize the Soul part of another—sometimes (perhaps not often enough) it is an instant recognition between two "souls" who come together.

So what can we do to develop the Soul part of us? To become conscious of our Soul Body?

This is daily work—the way in which we live our daily life, doing the important thing first for that day, but then always making the time for quietness, stillness, silence, to go within, to still the mind as much as possible, even if for five minutes which is a beginning. We can learn to extend that time to no longer than twenty minutes.

This, which is a listening to the Inner Self, and then being greatly aware of nurturing oneself, is most important. We need enough time to be in Nature, even if in the garden, to work with the energies of Nature—to allow the sunlight, or even the Light, to ray down upon us, to openly accept it, especially through the eyes, and to be very conscious of allowing out the Joy within our Heart Centre, for it IS there, sometimes very locked in. Laughter can help to release it.

This is all work towards that Peace which can be felt within. Then when a time comes which may promote a "ruffle" in the heart and solar plexus, we are ready and prepared to act compassionately, even to ourselves, and allow the calm waters of our Soul quality to take over.

This then is true Soul quality and awareness—first to be aware of oneself, one's own needs, and then those of others.

Releasing the Cobwebs

Cobwebs come and go, depending on how we cope with life's possibilities from day to day. We travel along life's Pathway, sometimes heeding challenges, sometimes getting onto the sidewalk to avoid what seem like pitfalls, and in so doing, we weave objections and every reason under the Sun why we should divert from that direction—like putting up a smoke screen around us. We don't want to see clearly any more, so the cobweb is very effective, and we divert our attention purposefully, making the cobweb strands even thicker. The Sun cannot penetrate properly, and leaves only shadows where the Light is really needed.

Now we have to work even harder to get through the mass of tiny threads, to sort them out and make a clearing to allow the Light to penetrate. This is very important work. We so encourage each Soul to begin that work now. Ask your Shining Ones to help you daily to begin to clear the shadows, to push aside the fluffy strands—examining them, seeing them for what they represent, and then with Love and understanding discarding. Slowly the Light becomes all there is.

This is Peace—this is true health.

MEDITATION: THE TEMPLE OF YOUR STAR

Come with me on a mission. . . .
Let us feel that we are ready for a journey into the Temple of your very Being. . . .
We are filled with Love, Compassion, Joy, Understanding. . . .
We have withstood many tests and now we are going forth
through unknown territory to our True Home. . . .

Feel the lightness in your body. . . .
Feel, as you begin to move upwards, the quality of Light within. . . .
Feel it shimmering . . . radiating . . . sending a Light before you. . . .
Your whole Being is filled with Love, which is Light . . .
and as you journey forwards and upwards within that Light,
feel that Oneness with all Being . . . all Nature . . .
all Space . . . stars . . . planets. . . .

Now you journey faster, swiftly carried on the tide of Light . . .
towards a great Star shining in the distance . . . YOUR Star . . . your Home . . .
and your radiance is drawn into its shining beams . . . enfolding you . . .
drawing you into the very centre of its heart . . .
there to commune with Shining Beings
who are your Brothers and Sisters of Time. . . .

Be at Peace in the Oneness of All in the Temple of your Star within. . . .

NOTES

1. An extract from Omraam Mikhaël Aïvanhov, *Light is a Living Spirit* (Fréjus, France: Editions Prosveta, 1983), 27–32, 62.
2. Alice Friend, article in *Stella Polaris* (New Lands, Brewells Lane, Liss, Hampshire, GU33 7HY U.K., The White Eagle Publishing Trust).
3. Ron Minson, M.D. "A Sonic Birth," in *Music and Miracles*, compiled by Don G. Campbell (Wheaton, IL: Quest Books, 1992) 91–92.
4. Hal Lingerman, *The Healing Energies of Music* (Wheaton, IL: Quest Books, 1983) 82.
5. Jan Linch, *Starchild* (Maidstone, Kent, U.K.: Haven Publishers, 1999) 6, 9.
6. David Wallechinsky & Amy Wallace, editors, *The Book of Lists—'90s Edition* (New York: Little, Brown & Co., 1993) 140.
7. From *Hands of Light* by Barbara Ann Brennan, copyright © 1987 by Barbara A. Brennan. Used by permission of Bantam Books, a division of Random House, Inc. 63–65.
8. Gladys McGarey, M.D., *Born to Live* (Santa Barbara, CA: Ross-Erikson Publishers, 1980) 53–57, 60.
9. Ibid., 82–83.
10. Suzy Chiazzari, *The Complete Book of Color* (Shaftesbury, Dorset, U.K.: Element Books, 1998) 155.
11. Ibid., 154.
12. Ibid., 154.
13. Excerpted from *Notes From the Song of Life*, copyright © 1977, 1987 by Tolbert McCarroll. Reprinted by permission of Celestial Arts, P.O. Box 7123, Berkeley, CA 94707, 27.
14. Lee Carroll and Jan Tober, *The Indigo Children* (Carlsbad, CA: Hay House, Inc., 1999) 1–2, 6–7, 10, 13, 15, 18, 21, 23, 30, 31.
15. Janet Hobbs, *The Gateway* (Langley, BC: JourNet Communications Ltd., 1997) 61–62.
16. Corinne Helene, *Healing and Regeneration Through Color* (Marina del Rey, CA: DeVorss & Company, 1983, 20th Printing) 62–63.
17. Tony Cooper, *International Association of Colour Newsletter (Winter 2000)*. Available from IAC, 46 Cottenham Road, Histon, Cambridge, CB4 9ES, U.K. Tel: 01223 563403; Email <iac@cix.co.uk>
18. An extract from Omraam Mikhaël Aïvanhov, *Light is a Living Spirit* (Fréjus, France: Editions Prosveta, 1983) 62, 67.
19. Ronald P. Beesley, *The Creative Ethers* (London, U.K.: Neville Spearman Limited, 1975) 99, 101, 105.
20. Sheldan Nidle, *Your First Contact* (Pukalani, HI: Blue Lodge Press, 2000) 69.
21. Anodea Judith, *The Truth About Chakras* (St. Paul, MN: Llewellyn Publications, 1999) 6–7.
22. Jan Linch, *Starchild* (Maidstone, Kent, U.K.: Haven Publishers, 1999) 18.
23. Ibid., 16.
24. Mitchell Gaynor, M.D., *Healthy Living Magazine (July/August 1999)*.
25. Dr. Robert D. Willix, Jnr., *Health & Longevity Newsletter (June 1996)* 5. Available from: #306 - 1515 S. Federal Highway, Boca Raton, FL 33432.
26. Corinne Heline, *Music: The Keynote of Human Evolution* (La Canada, CA: New Age Press, 1965) 120.
27. Vera Stanley Alder, *The Finding of the Third Eye* (York Beach, ME: Samuel Weiser, Inc., 1973) 107. Material used by permission.
28. William David, *The Harmonics of Sound, Color & Vibration* (Marina del Rey, CA: DeVorss & Company, 1980) 65.
29. Corinne Heline, *Music: The Keynote of Human Evolution* (La Canada, CA: New Age Press, 1965) 106.
30. Ted Andrews, *The Healer's Manual* (St. Paul, MN: Llewellyn Publications, 1993) 152.
31. Laurel Elizabeth Keyes, *Toning: The Creative Power of the Voice* (Marina del Rey, CA: DeVorss & Company, 1973).
32. Olivea Dewhurst-Maddock, *The Book of Sound Therapy: Heal Yourself with Music and Voice* (New York: Simon & Schuster, Inc., 1993) 111.
33. Excerpt from a transcript of a talk given by Awahoshi Kavan, B.A. Communication Disorder, M.A. Instructional Technology, at the Mind/Body/Spirit Conference, London, U.K., in 1998, *Whole Life News (September 1998)*, Namaste Publishing, 5. Awahoshi Kavan is a pioneer of crystal sound. Chrysalis, Piazza Federico Di Svevia, 95, 95121 Catania, Italy. Website: www.bali3000.com/crystalsound
34. Betina Lindsey, *Creativity Links (February 2000)* Springdale, UT.
35. Corinne Heline, *Music: The Keynote of Human Evolution*, (La Canada, CA: New Age Press, 1965) 140.
36. Excerpted from *Notes From the Song of Life*, copyright © 1977, 1987 by Tolbert McCarroll. Reprinted by permission of Celestial Arts, P.O. Box 7123, Berkeley, CA 94707, 16.

37. Dr. Robert D. Willix, Jnr., *Health & Longevity Newsletter (June 1996)* 5. Available from: #306 - 1515 S. Federal Highway, Boca Raton, FL 33432.
38. Ibid, 5.
39. Ronald P. Beesley, *Union of Energies: Galilee Series I* (Speldhurst, Kent, U.K.: White Lodge Publications, 1974) 56.
40. Steven Halpern, from his website *Steven Halpern's Inner Peace Music* at www.innerpeacemusic.com/sound_healing.html
41. Jan Linch, *Starchild* (Maidstone, Kent, U.K., Haven Publishers, 1999) 14.
42. Jule Klotter, "Sound Therapy," *Townsend Letter for Doctors & Patients (July 1997)* 12–16.
43. John Diamond, M.D., *Your Body Doesn't Lie* (New York: Warner Books, 1979) 159–160.
44. Jean Houston was interviewed for the article "Sound and the Miraculous" in *Music and Miracles*, compiled by Don G. Campbell (Wheaton, IL: Quest Books, 1992) 15.
45. Awahoshi Kavan, "Pure Tone Crystal Music Therapy" in *Whole Life News (September 1998)*, Namaste Publishing, 13.
46. Sky Holden Dunn, in brochure by Crystal Tones, distributor of crystal bowls in Salt Lake City, UT. Tel: 1-800-358-9492.
47. Awahoshi Kavan, "Pure Tone Crystal Music Therapy" in *Whole Life News (September 1998)*, Namaste Publishing, 13.
48. Fabien Maman, The Academy of Sound, Color and Movement, 4800 Baseline #E104-237, Boulder, CO 80303. Tel 1-800-615-3675. Fax (303) 926-0552. Email info@tama-do.com or tamadoacademy@aol.com
49. Jeanne Elizabeth Blum, *Woman Heal Thyself* (Boston: Charles E. Tuttle Co. Inc., 1995) 254–255.
50. Suzy Chiazzari, *The Complete Book of Color* (Shaftesbury, Dorset, U.K.: Element Books Limited, 1998) 109.
51. White Eagle, *Stella Polaris (April/May 1996)* 88–89.
52. Annie Wilson and Lilla Bek, *What Colour Are You?* (Wellingborough, Northamptonshire, U.K.: Turnstone Press Limited, 1981) 35.
53. Alice Bailey, *Letters on Occult Meditation* (New York: Lucis Publishing Company, Fifteenth Printing 1993) 247.
54. An extract from Omraam Mikhaël Aïvanhov, *Toward a Solar Civilisation* (Fréjus, France: Editions Prosveta, 1982) 83–84.
55. Lucy Hawking, *Daily Mail*, January 17, 1995.
56. Audrey Ann Lowrie, *Colour Therapy, Part I, Home Study Course* (Perkinsfield, ON: The Colour Institute of Canada, 1993) 51.
57. Marie Louise Lacy, *Know Yourself Through Colour* (London, U.K.: HarperCollins Publishers, 1989) 13.
58. Annie Wilson and Lilla Bek, *What Colour Are You?* (Wellingborough, Northamptonshire, U.K.: Turnstone Press Limited, 1981) 38–39.
59. Ibid., 20.
60. From *Hands of Light* by Barbara Ann Brennan, copyright © 1987 by Barbara A. Brennan. Used by permission of Bantam Books, a division of Random House, Inc. 269.
61. Janice Ellicott, letter to the editor in *International Journal of Alternative & Complementary Medicine (May 1995)*.
62. Audrey Ann Lowrie, *Colour Therapy, Part I, Home Study Course* (Perkinsfield, ON: The Colour Institute of Canada, 1993) 64.
63. Paula Marie, letter to the editor in *International Journal of Alternative & Complementary Medicine (May 1995)*.
64. Annie Wilson and Lilla Bek, *What Colour Are You?* (Wellingborough, Northamptonshire, U.K.: Turnstone Press Limited, 1981) 20–21.
65. *Light: Medicine of the Future* by Jacob Liberman, O.D., Ph.D., published by Bear & Company, Rochester, VT 05767 Copyright © 1991 by Jacob Liberman, O.D., Ph.D. 91, 99.
66. Ronald P. Beesley, *The Creative Ethers* (London, U.K.: Neville Spearman Limited) 96–97.
67. From *Hands of Light* by Barbara Ann Brennan, copyright © 1987 by Barbara A. Brennan. Used by permission of Bantam Books, a division of Random House, Inc.
68. Jeanne Elizabeth Blum, *Woman Heal Thyself* (Boston: Charles E. Tuttle Co. Inc., 1995).
69. Fabien Maman, *Healing with Sound, Color and Movement, Book IV* (CA: Tama-Dõ Press, 1997) 69.
70. From *Hands of Light* by Barbara Ann Brennan, copyright © 1987 by Barbara A. Brennan. Used by permission of Bantam Books, a division of Random House, Inc. 69–70.
71. Mitchell Gaynor, M.D., *Sounds of Healing* (New York: Broadway Books, 1999) 24.

Bibliography

Alder, Vera Stanley. *The Finding of the Third Eye*. York Beach, ME: Samuel Weiser, Inc., 1973.

Aïvanhov, Omraam Mikhaël. *Light is a Living Spirit*. Fréjus, France: Editions Prosveta, 1983.

———. *Man's Subtle Bodies and Centres*. Fréjus, France: Editions Prosveta, 1986.

———. *Toward a Solar Civilisation*. Fréjus, France: Editions Prosveta, 1982.

Andrews, Ted. *The Healer's Manual*, St. Paul, MN: Llewellyn Publications, 1993.

Bailey, Alice. *Letters on Occult Meditation*. New York: Lucis Publishing Company, Fifteenth Printing 1993.

Beesley, Ronald P. *The Creative Ethers*. London, U.K.: Neville Spearman Limited, 1975.

———. *Union of Energies: Galilee Series I*. Speldhurst, Kent, U.K.: White Lodge Publications, 1974.

Blum, Jeanne Elizabeth. *Woman Heal Thyself*. Boston: Charles E. Tuttle Co., Inc., 1995.

Brennan, Barbara Ann. *Hands of Light: A Guide to Healing Through the Human Energy Field*. New York: Bantam New Age, 1987.

Brodie, Renee. *Life Energies*. Vancouver, BC: Vancouver White Lodge Association, 1984.

———. *The Healing Tones of Crystal Bowls: Heal Yourself with Sound and Colour*. Delta, BC: Aroma Art Ltd., Third Printing 2000.

Campbell, Don G., compiler. *Music and Miracles*. Wheaton, IL: Quest Books, 1992.

Carroll, Lee and Jan Tober. *The Indigo Children*. Carlsbad, CA: Hay House, Inc., 1999, 10th Printing 2000.

Chiazzari, Suzy. *The Complete Book of Color*. Shaftesbury, Dorset, U.K.: Element Books Limited, 1998.

Cornell, Judith, PhD. *Drawing the Light from Within*. New York: Prentice Hall Press, 1990.

David, William. *The Harmonics of Sound, Color & Vibration*. Marina del Rey, CA: DeVorss & Company, 1980.

Dewhurst-Maddock, Olivea. *The Book of Sound Therapy: Heal Yourself with Music and Voice*. New York: Simon & Schuster Inc., 1993.

Diamond, John, M.D. *Your Body Doesn't Lie*. New York: Warner Books, 1979.

Dinshah, Darius. *Let There Be Light*. Malaga, NJ: Dinshah Health Society, 1985.

Gaynor, Mitchell, M.D. *Sounds of Healing*, New York: Broadway Books, 1999.

Goldman, Jonathan. *Healing Sounds*. Rockport, MA: Element Books Limited, 1992.

Halpern, Steven. *Sound Health: The Music and Sounds that Make Us Whole*. San Francisco: Harper & Row, 1985.

Heline, Corinne. *Color and Music in the New Age*. Marina del Rey, CA: Devorss & Company, 1964.

———. *Healing and Regeneration Through Color / Healing and Regeneration Through Music*. Marina del Rey, CA: DeVorss & Company, 1983.

———. *Music: The Keynote of Human Evolution*. La Canada, CA: New Age Press, 1965.

Hobbs, Janet. *The Gateway*. Langley, BC: JourNet Communications, Ltd., 1997.

Judith, Anodea. *The Truth About Chakras*. St. Paul, MN: Llewellyn Publications, 1999.

Keyes, Laurel Elizabeth. *Toning: The Creative Power of the Voice*. Marina del Rey, CA: DeVorss & Company, 1973.

Lacy, Marie Louise. *Know Yourself Through Colour*. London, U.K.: HarperCollins Publishers, 1989.

Liberman, Jacob. *Light: Medicine of the Future*. Santa Fe, NM: Bear & Company Publishing, 1991.

Linch, Jan. *Starchild*. Maidstone, Kent, U.K.: Haven Publishers, 1999.

Lindsey, Betina. Article in *Creativity Links*, February 2000. Mailing address: P.O. Box 362, Springdale, UT 84767. Tel: 1-801-295-8051.

Lingerman, Hal. *The Healing Energies of Music*. Wheaton, IL: Quest Books, 1983.

Lowrie, Audrey Ann. *Colour Therapy, Part I, Home Study Course*. Perkinsfield, ON: The Colour Institute of Canada, 1993. Tel (705) 549-1999.

Maman, Fabien. *Healing with Sound, Color and Movement, Book IV*. CA: Tama-Dõ Press, 1997 (P.O. Box 7000-746, CA 90277).

McCarroll, Tolbert. *Notes From the Song of Life*. Berkeley, CA: Celestial Arts, 1977.

McGarey, Gladys, M.D. *Born to Live*. Santa Barbara, CA: Ross-Erikson Publishers, 1980.

Nidle, Sheldan. *Your First Contact*. Pukalani, HI: Blue Lodge Press, 2000.

Tame, David. *The Secret Power of Music*. New York: Destiny Books, 1984.

White Eagle. *Stella Polaris, April/May 1996*.

Willix, Dr. Robert D., Jnr. Article in *Health & Longevity Newsletter (June 1996)*. Mailing address: Suite 306, 1515 South Federal Highway, Boca Raton, FL 33432.

Wilson, Annie and Lilla Bek. *What Colour Are You?* Wellingborough, Northamptonshire, U.K.: Turnstone Press Limited, 1981.